Azeotropes
Behavior & Application

by D. James Benton

Forward

Azeotrope refers to a mixture of two or more fluids whose proportions cannot be altered by simple distillation. Azeotropes exhibit the same mass fraction of constituents in the vapor and liquid phases; thus, when boiled, the vapor has the same composition as the liquid. This behavior differs from ideal solutions, where one component is typically more volatile than the others. The azeotrope may have a higher or lower boiling point than the constituents. We study these anomalous fluids because they may exhibit properties that are more advantageous than the constituents. A similar motivation accompanies multi-weight lubricants. In this text we will explore thermodynamic and transport properties as well as applications, including: refrigeration and vapor power cycles.

All of the examples contained in this book,
(as well as a lot of free programs) are available at...
https://www.dudleybenton.altervista.org/software/index.html

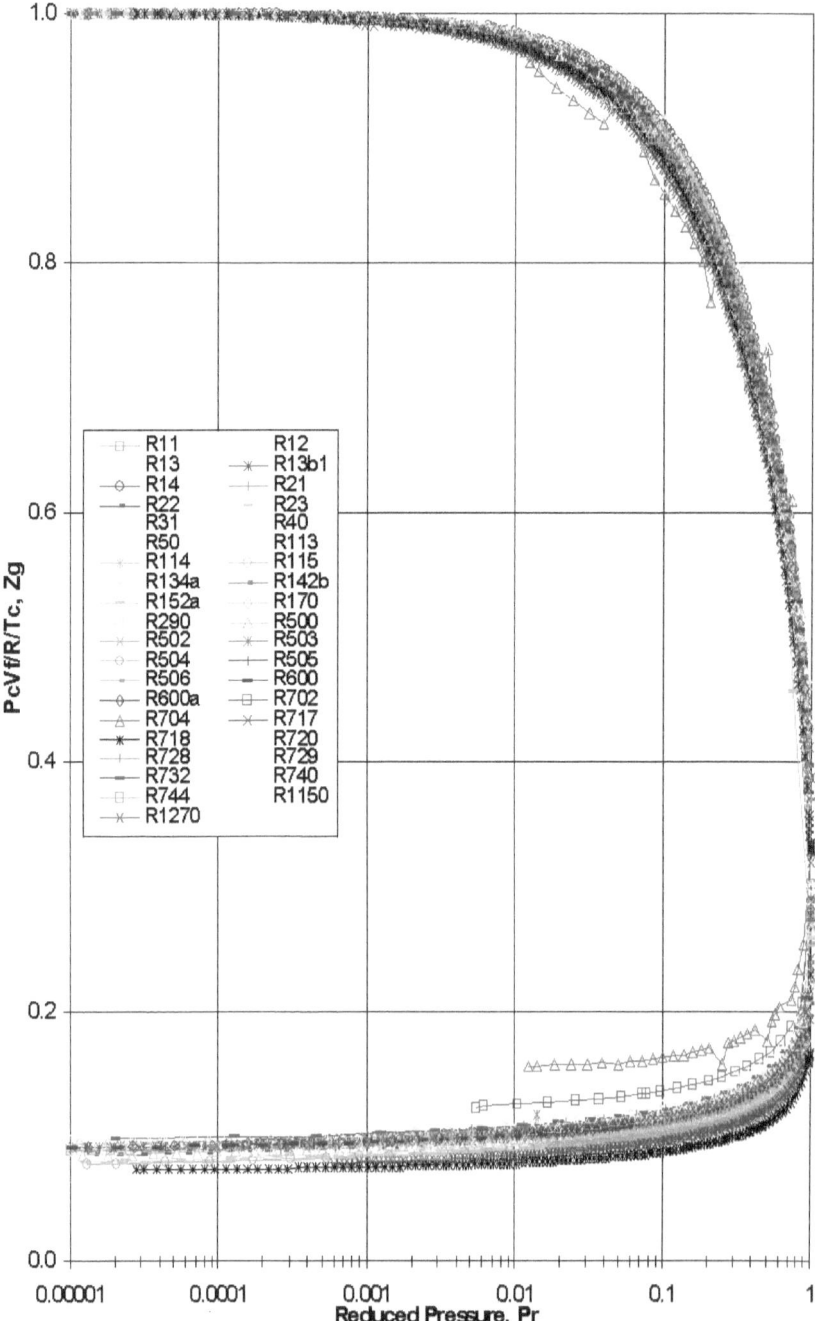

Table of Contents page

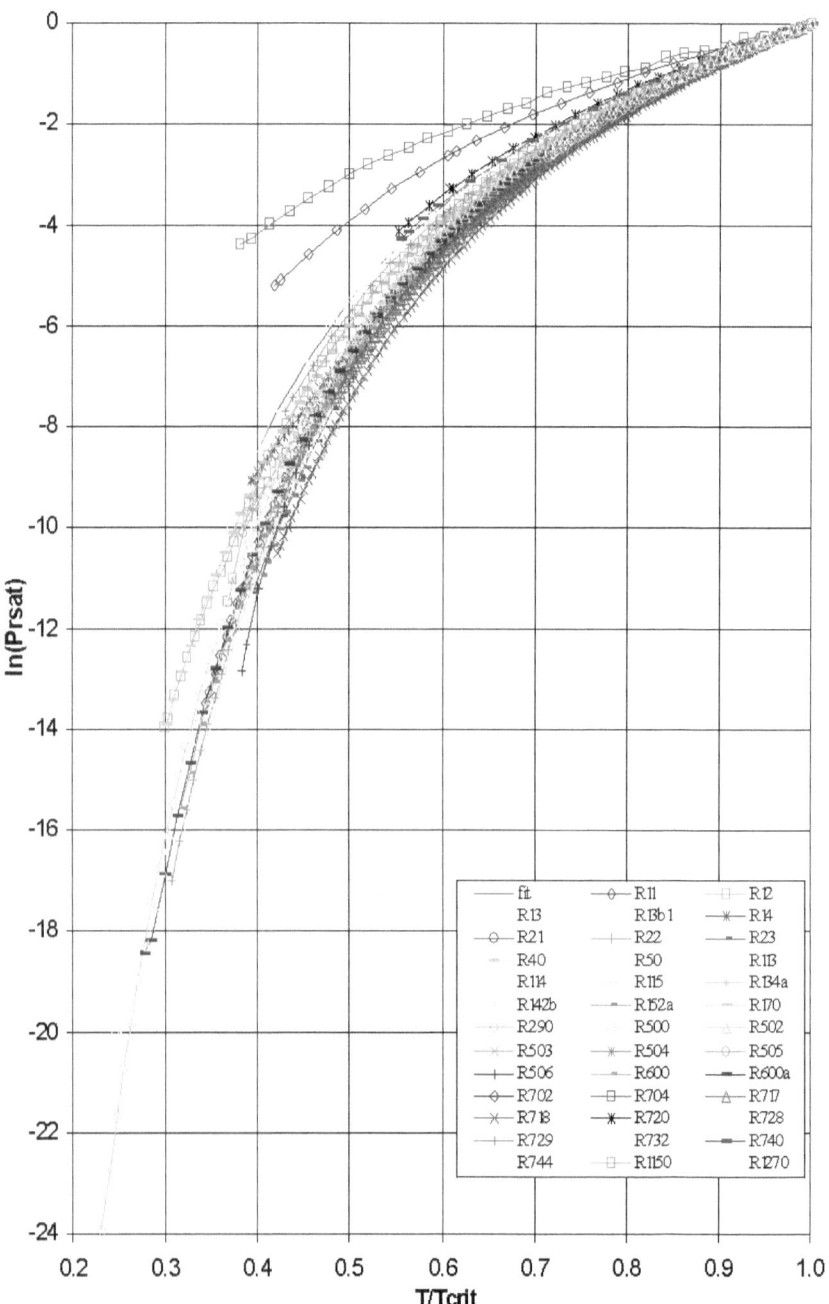

Chapter 1. Introduction

Azeotropes do not follow Raoult's law; thus, we begin with this important discovery. Raoult[1] found that the partial vapor pressure of each constituent in a mixture was proportional to the vapor pressure of the pure substance times the mole fraction. As this seemed to apply for a variety of mixtures, he presumed it to be a *law*. As we now know this does not hold for all mixtures, it might be termed a *relationship*. This is similar to Dalton's "law" of partial pressures.[2]

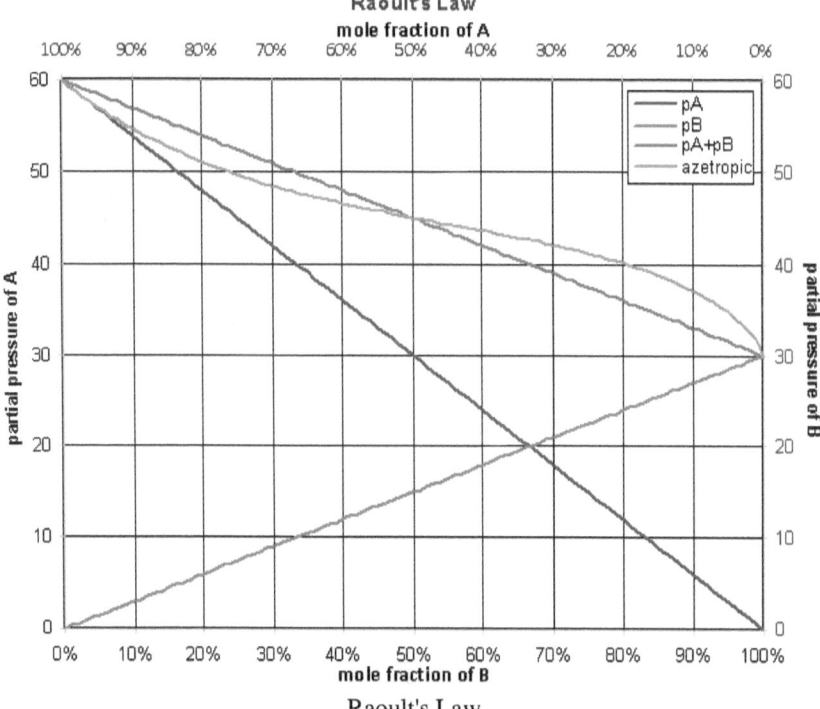

Raoult's Law

Raoult's Law is illustrated in the figure above. Ideal behavior would follow the downward-sloping green line, while azeotropic behavior would follow the orange curve or something similar, but not the straight green line. Considerable effort has been invested in measuring the behavior of various mixtures in the laboratory, not only to better understand the science but also to seek out new and interesting behavior, which might prove to be advantageous in certain applications, including refrigeration and power cycles. This data is also of

---

[1] François-Marie Raoult (1830–1901) French research chemist who studied solutions.
[2] In a mixture of non-reacting gases, the total pressure exerted is equal to the sum of the partial pressures of the individual gases, which are proportional to the mole fractions. John Dalton (1766–1844) English chemist, physicist, and meteorologist.

1

interest to the refinery and chemical industry, as it relates to the separation of mixtures or the difficulty thereof. One remarkable example comes from mixing water and 2-propanol. Data from Brunjes and Bogart[3] are shown in the following figure:

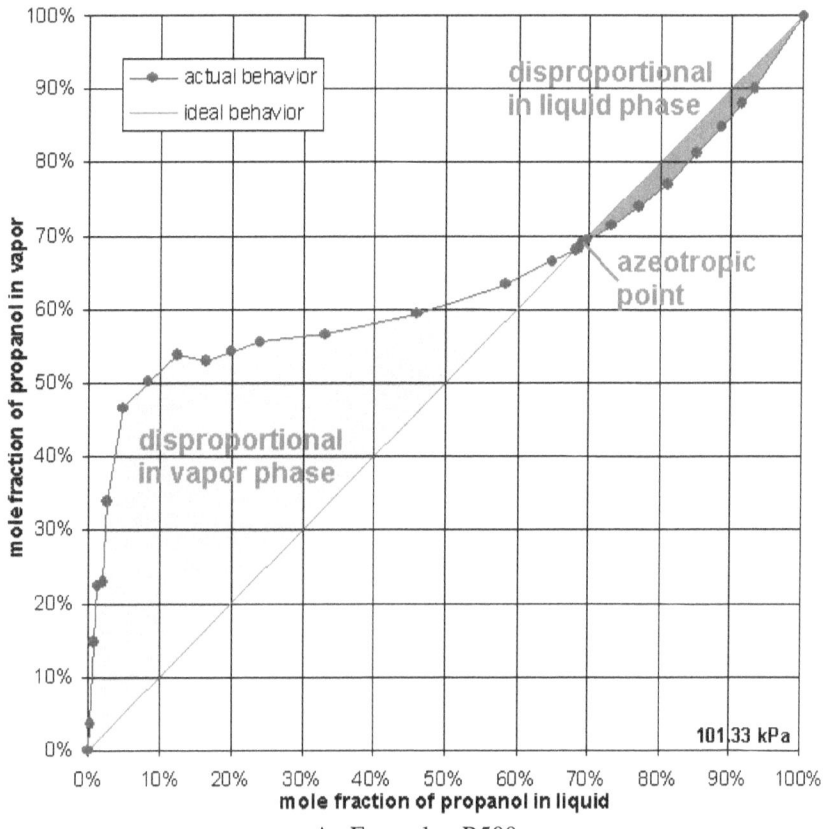

An Example - R500

Refrigerant 500 has been in use for some time and was one of the early available azeotropes. The *ASHRAE Handbook of Fundamentals* is perhaps the seminal resource for such properties. R500 is a 73.8/26.2 mixture (by mass) of R12 and R152a. The mole fractions are 60.6/39.4. A comparison of measured and calculated critical properties are listed in the following table.

[3] Brunjes, A. S. and Bogart M. J. P., "The Binary Systems Ethanol-n-Butanol, Acetone-Water and Isopropanol-Water," Industrial & Engineering Chemistry, Vol. 35, pp. 255-260, 1943.

comparison		constituents		R500		
property	units	R12	R152a	ideal	actual	difference
MW	g/mole	120.93	66.05	99.30	99.30	N/A
Tc	°K	385.2	385.4	385.5	375.3	2.7%
Pc	kPa	4115	4520	4222	4173	1.2%
Vc	m³/kg	0.001792	0.002717	0.002034	0.002012	1.1%
Zc	-	0.2785	0.2525	0.2717	0.2672	1.7%
Hc	kJ/kg	183.4	314.4	237.8	210.7	3.4%
Sc	kJ/kg/°K	0.5690	0.9594	0.6713	0.6609	1.6%

mole ratio 60.6%   39.4%

mass ratio 73.8%   26.2%

The data may be found in spreadsheet R500.xls in the examples folder of the online archive. The pressure-volume behavior is illustrated in the following figure:

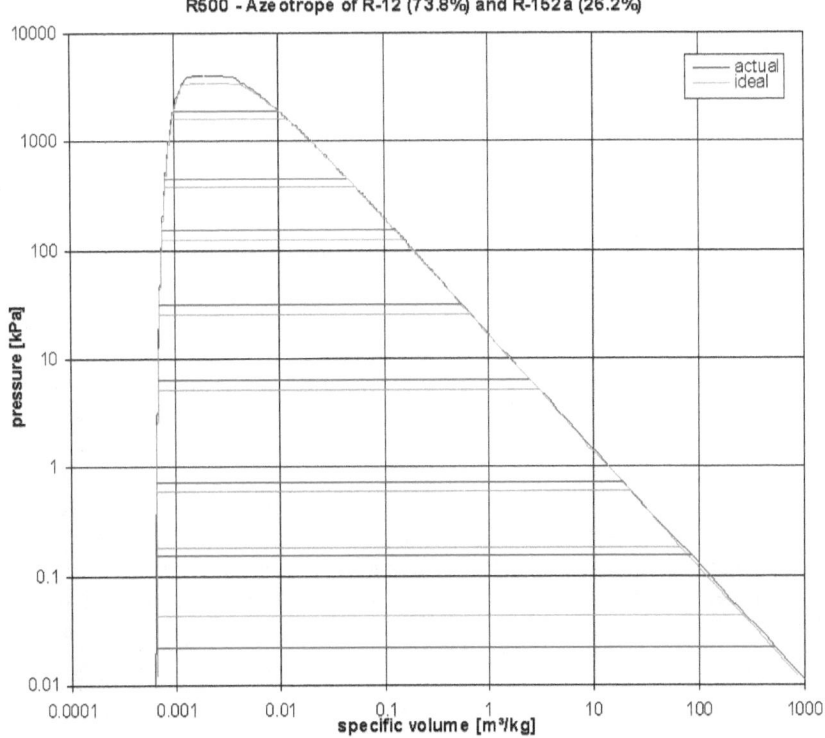

R500 - Azeotrope of R-12 (73.8%) and R-152a (26.2%)

The ideal (orange) lines are quite close to the actual (blue) lines, indicating that the measured behavior somewhat follows the calculated, based on the ideal assumption. This is a log-log plot, so the differences may be visually

3

underestimated. The next figure shows enthalpy vs. entropy, often called a Mollier diagram and is vital to refrigeration and power cycles.

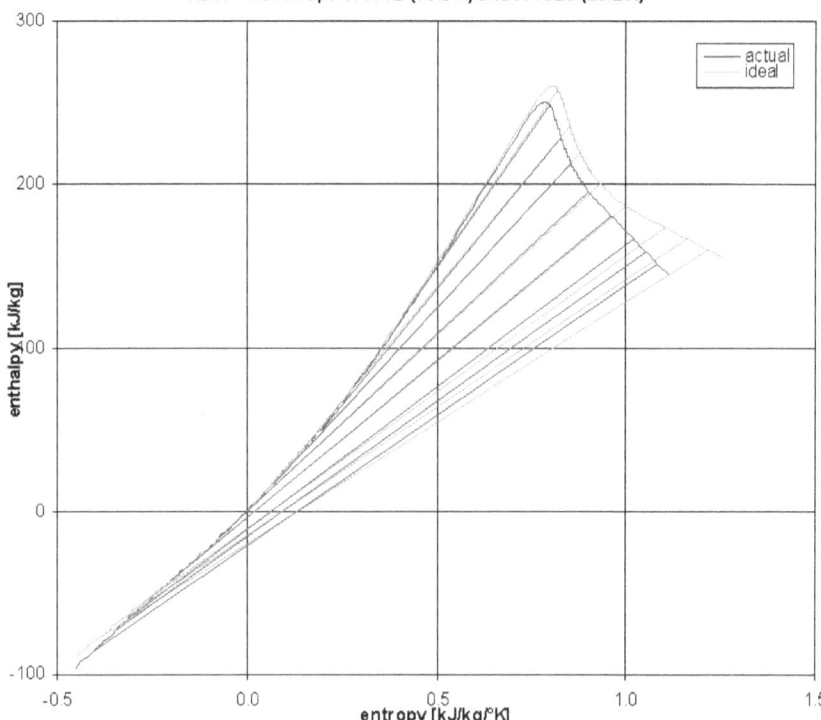

This figure has linear axes and shows the greatest differences in the vapor state (upper right corner). The departure (actual from ideal) is more pronounced at lower vapor pressures than at higher (near 1.2:160 than 0.8:250). The lower pressure conditions would be seen at the exit of the evaporator and inlet to the compressor for a refrigeration cycle. While the molecular weight of R12 is almost twice that of R152a, the critical temperatures differ by only 0.2°K and the critical pressures differ by only 10%, making R500 a rather boring azeotrope. In order to see remarkable differences, we need to examine a mixture of unlike fluids.

An unusual and interesting graph which covers the entire range of properties on a linear scale is $T_R(1-Z)$ vs. V_C/V. Unlike the previous graphs, this one is continuous in slope even across the vapor dome. The critical point is not at the top; rather, it is at $V_C/V=1$ and $1-Z_C$.

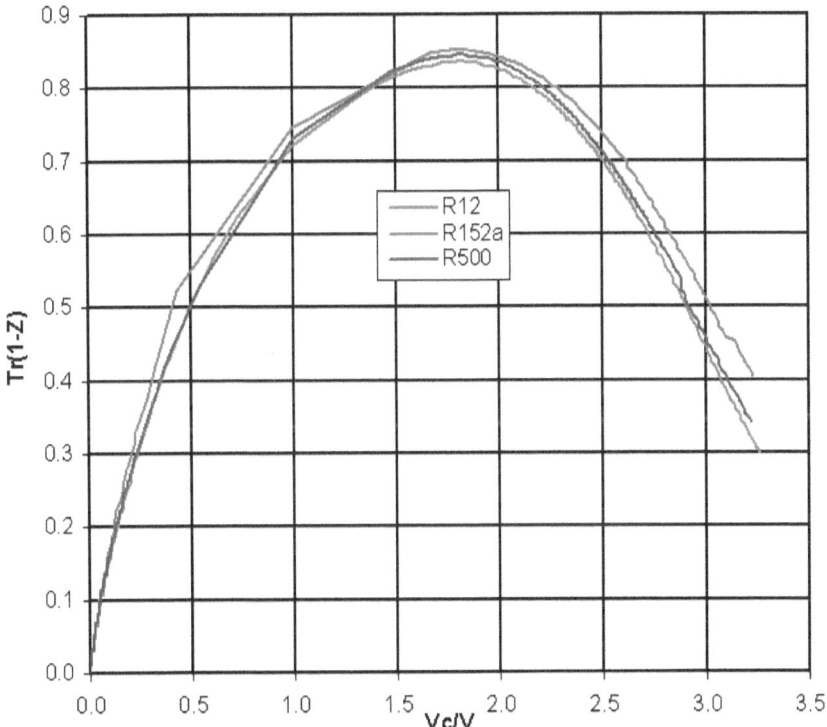

Chapter 2. Azeotropic Behavior

Of the common two-component azeotropic refrigerants (i.e., 400 and 500 series), R506 has the most differing constituents. R506 is a mixture of R31 and R114 with a 55.1/44.9 ratio by mass. Properties of the three fluids and ideal mixture are summarized in the following table:

comparison		constituents		R506		
property	units	R31	R114	ideal	actual	difference
MW	g/mole	68.48	170.92	93.70	93.70	N/A
Tc	°K	424.9	418.8	422.2	414.8	1.8%
Pc	kPa	5131	3257	4289	5167	-17.0%
Vc	m³/kg	0.002315	0.001724	0.002049	0.001815	12.9%
Zc	-	0.2302	0.2756	0.2506	0.2345	6.9%
Hc	kJ/kg	323.0	221.1	277.2	261.0	6.2%
Sc	kJ/kg/°K	0.8859	0.6621	0.7854	0.7687	2.2%

mole ratio 75.4%   24.6%

mass ratio 55.1%   44.9%

Dichlorotetrafluoroethane (R114) is an unremarkable refrigerant, while Chlorofluoromethane (R31) is quite remarkable, producing interesting properties of the azeotrope. The critical compressibility, Z_C, of R31 alerts us that this is an unusual fluid. The average critical compressibility of 36 common refrigerants is 0.2675, well above that for R31, which is 0.2302. (See Appendix A for more critical properties and Appendix B for complete properties of R31.) While this difference might not seem too large, consider the 9 most similar common fluids listed in the following table:

name	formula	MW	Tc	Pc	Vc	Zc
		-	°K	kPa	m³/kg	-
Methanol	CH3OH	32.04	513.0	8090	0.00313	0.1900
Lithium	Li	6.94	3220.0	66100	0.01225	0.2100
Cesium	Cs	132.91	1938.0	9500	0.00268	0.2100
Potassium	K	39.10	2223.0	15999	0.00632	0.2137
Water	H2O	18.02	647.1	22064	0.00311	0.2294
Sodium	Na	22.99	2573.0	34998	0.00612	0.2300
Chlorofluoromethane	CH2FCl	68.48	424.9	5131	0.00231	0.2302
Nitric Acid	HNO3	63.01	520.0	6890	0.00230	0.2311
Ammonia	NH3	17.03	405.5	11356	0.004251	0.2439
Ethanol	C2H5OH	46.07	514.0	6300	0.00365	0.2480

With the exception of the liquid metals (Lithium, Cesium, Potassium, and Sodium), all of these fluids are *polar*, which means the charge is very much different on one side of the molecule than the other. The polar compounds also have high bond stresses. Water is an extraordinary fluid. Ammonia is the most outstanding of all refrigerants, having the highest latent heat, which means the lowest flow rate to achieve a given refrigeration load. This is why old ice factories and skating rinks always used Ammonia refrigeration systems. Some still do. Ammonia has serious safety issues and is also used to create illegal drugs, so its use is rapidly disappearing. Methanol and Ethanol are also extraordinary, being small hydrocarbons with a hydroxyl (OH) group. These also have very high latent heats (it takes an unusual amount of energy to evaporate or condense them). Right in the middle of these liquid metals and exceptional polar fluids is R31. It should not be surprising that it too exhibits unusual behavior. The compressibility of R31, R114, and H2O are compared in this next figure:

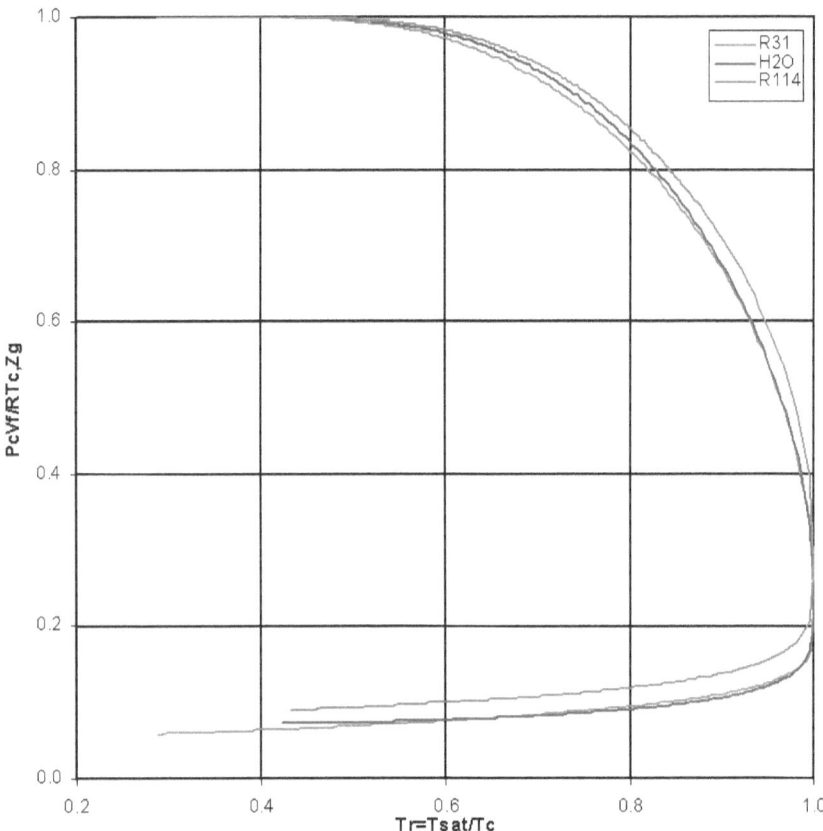

The horizontal axis is the reduced temperature, $T_R=T/T_C$. The vertical axis is both $Z_C V_F/V_C$ and $P_S V_G/RT_S$. All three curves touch the right side ($T_R=1$) at $Z=Z_C$. The upper portion of the curves is the compressibility of the saturated vapor. The lower portion of the curves is the reduced specific volume of the saturated liquid times the critical compressibility. We see here that R31 behaves

8

more like water than it does R114. Of course, the absolute temperatures of the R31 are 220°K colder than for those of water, making it suitable for refrigeration applications. Comparing the right column of the tables on pages 3 (R500) and 5 (R506), we see that a simple mass-weighted average is inadequate to predict the behavior of R506. Ideal (mass-weighted average) and actual P-V behavior is shown in this next figure:

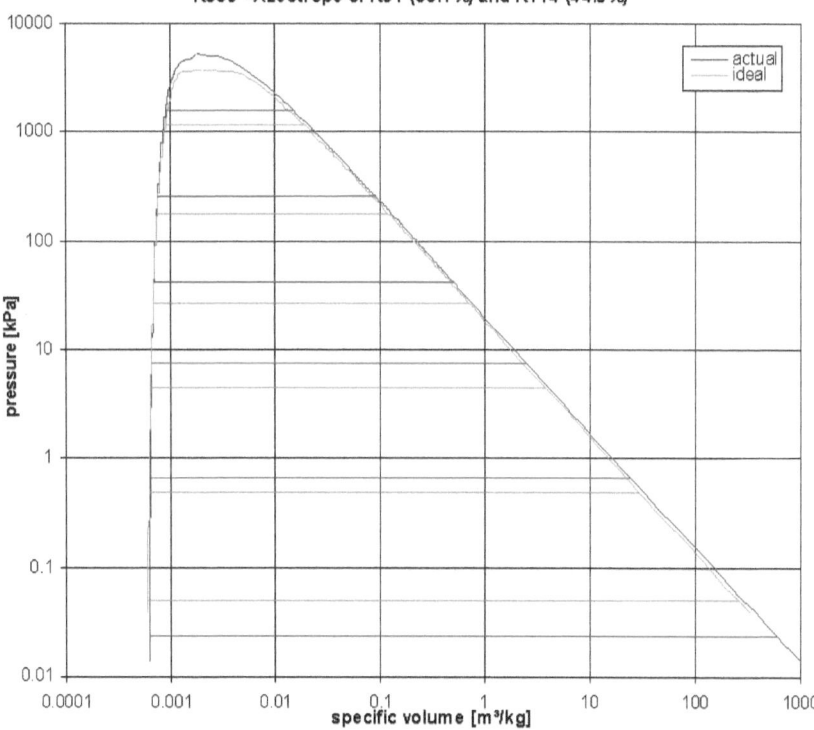

This figure is unremarkable and quite similar to the previous one for R500. The mass-weighted average under-predicts the saturation pressure at high temperatures (orange horizontal lines below blue) and under-predicts the saturation pressure at low temperatures (lowest orange horizontal line above blue). The Mollier diagram (H-S) for this azeotrope shows some discrepancies between ideal and actual, but is also unremarkable. The P-V diagram is useful for mechanical equipment sizing, especially compressors (in the case of refrigeration) or expanders (in the case of a power cycle). The Mollier diagram is most informative about the process inside the compressor or expander, which is why this chart is so often associated with steam turbines. The T-S diagram is most often associated with cycle efficiency and the H-P diagram is almost exclusively associated with refrigeration cycles. We will consider each of these here, starting with the Mollier.

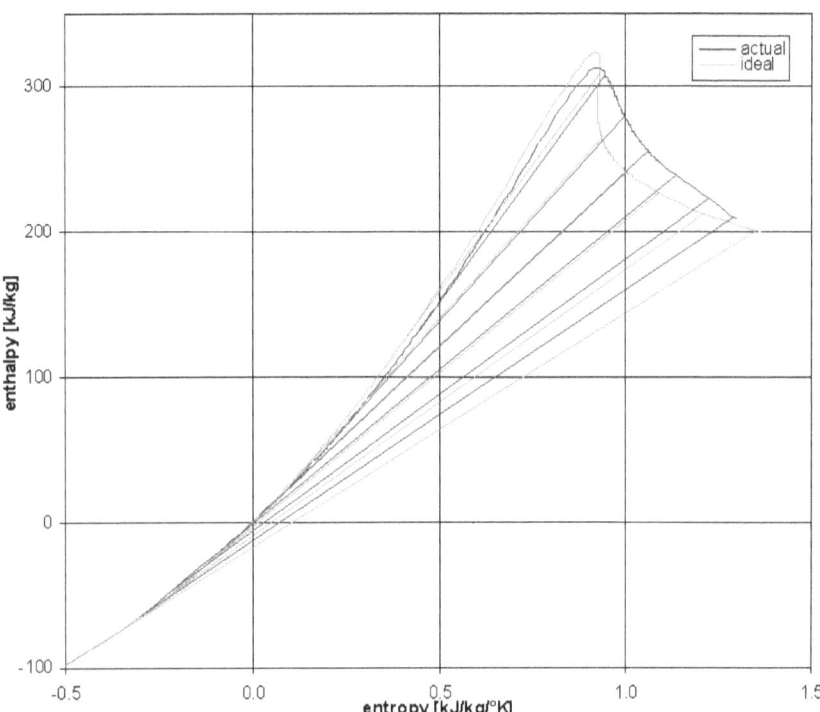

R506 - Azeotrope of R31 (55.1%) and R114 (44.9%)

The Carnot cycle is represented by a rectangle on a T-S diagram. This is the theoretical best attainable efficiency. Of course, a real refrigeration cycle must follow the actual process lines, including positive changes in entropy for any process other than negative heat transfer (out of the system). Still, the temperatures of the two reservoirs limits the achievable efficiency, as indicated by Carnot. We see deviations in the liquid (left side) and vapor (right side) regions of the vapor dome. A mass-weighted average under-estimates the entropy over most of the liquid/vapor region (orange curves to the left of the blue). This is consistent with the H-S diagram, at least in the upper right where it is visible. The bottom left side of a Mollier diagram is always squashed, which is why this region is often excluded. Note that, unlike every other thermodynamic property chart, lines on the Mollier diagram are continuous in slope. This unique feature is another reason for displaying processes on entropy-enthalpy axes.

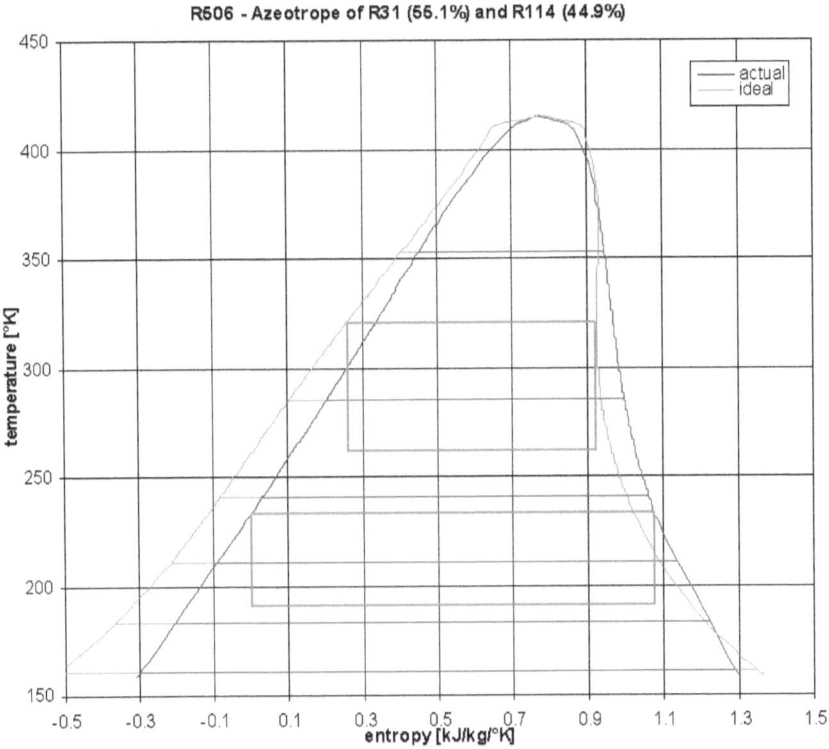

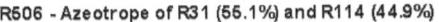

The diagonal sections of actual process lines deviate from the ideal rectangular shape of the Carnot cycle. Each of these deviations decreases the attainable efficiency, which is why a working fluid with less curvature along the vapor dome in the operating region is more advantageous purely from a thermodynamic efficiency perspective, although this same fluid might not be as desirable from other considerations (e.g., lower latent heat necessitating higher mass flow rate).

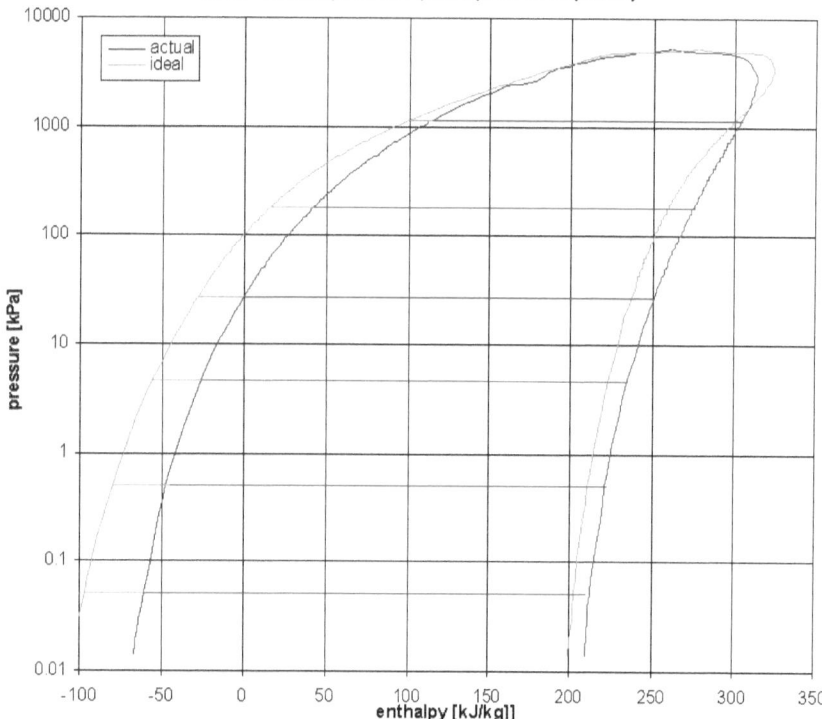

R506 - Azeotrope of R31 (55.1%) and R114 (44.9%)

The mass-weighted average also under-predicts the enthalpy, as shown in the figure above (orange curve left of blue). The latent heat ($H_{FG}=H_G-H_F$) is equal to the width of the vapor dome in the figure above. We see that the mass-weighted average over-predicts the latent heat (orange curve on the left side is farther to the left than the orange curve on the right side, making the length of the orange horizontal sections longer than the blue ones). As latent heat is key in either refrigeration or power cycles, the importance of accurate property formulations (rather than simplistic mixing rules) is shown here. These figures, along with the supporting data, can be found in the online archive (azeotropes.zip) in folder examples in spreadsheet R506.xls.

RKS Approximation

As the mass-weighted average does not yield sufficient accuracy, we next consider the Redlich-Kwong-Soave Equation of State (RKS EoS). Details of this useful relationship can be found in Appendix C and also in spreadsheet RKS.xls. This is a two-parameter cubic EoS. We match the liquid density and also Maxwell's Criterion to find the parameters A and B, which best fit the data for all three fluids. We could use the liquid and vapor densities, ignoring Maxwell's Criterion, but this would ultimately yield the wrong free energies and latent heats. The saturated vapor is much closer to the ideal gas state than the liquid, so we relax the condition of matching the saturated vapor specific volume in favor of the other two, as we have only two adjustable parameters. The first parameter, A, represents the van der Waals force (attractive) and is shown in this next figure:

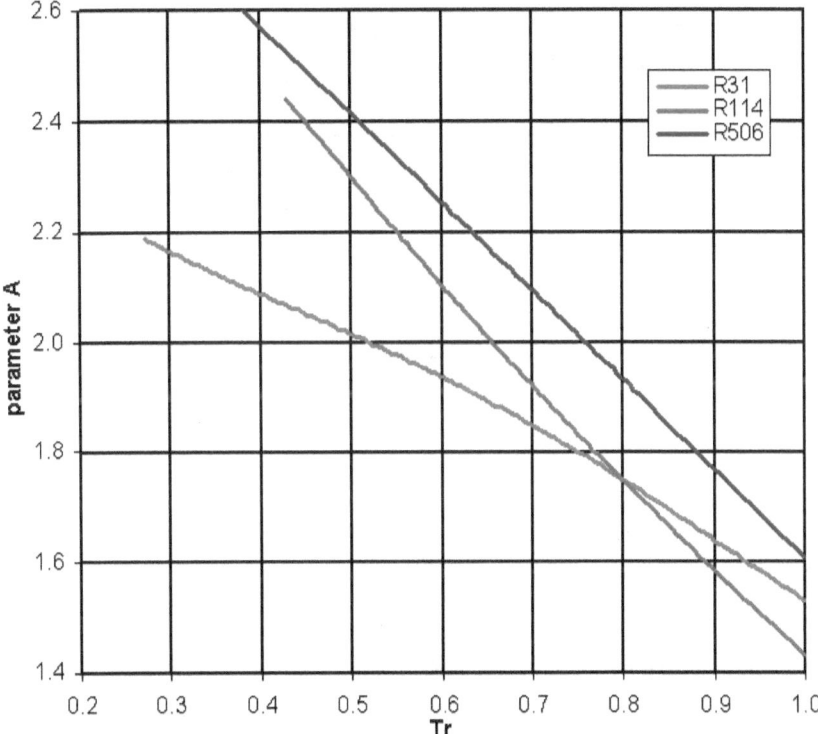

No combination of the red (R31) and green (R114) lines will yield the blue (R506) line. If the blue line lay between the red and green lines, then one might expect to predict the behavior of the azeotrope from its constituents, but this is not the case with this information alone or with this EoS, although which or both is not clear from what we have seen so far. Because the blue line is above the red and green, we conclude that the van der Waals (attractive) force is stronger for the azeotrope than either of its constituents.

The second parameter, B, represents the hard sphere volume (repulsive force) and is shown in this next figure:

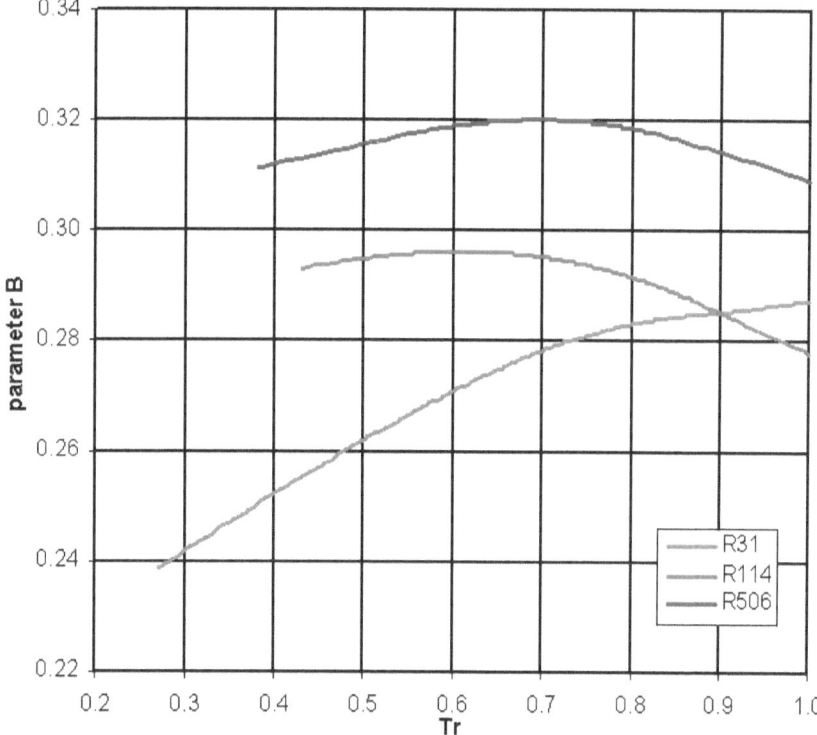

The same can be said of parameter B in that the red (R31) and green (R114) curves not indicative of the blue (R506) curve. Because the blue curve is above the red and green, we conclude that the hard sphere (repulsive force) is also stronger for the azeotrope than either of its constituents. Although we can't predict it from these parameters alone, the combination of two opposing forces explains the azeotropic behavior. The two differing molecules must fit together in a more complex way than they do with each other.

The plot of $T_R(1-Z)$ vs. V_C/V also shows blue curve outside the red and green, at least for the liquid range (right side).

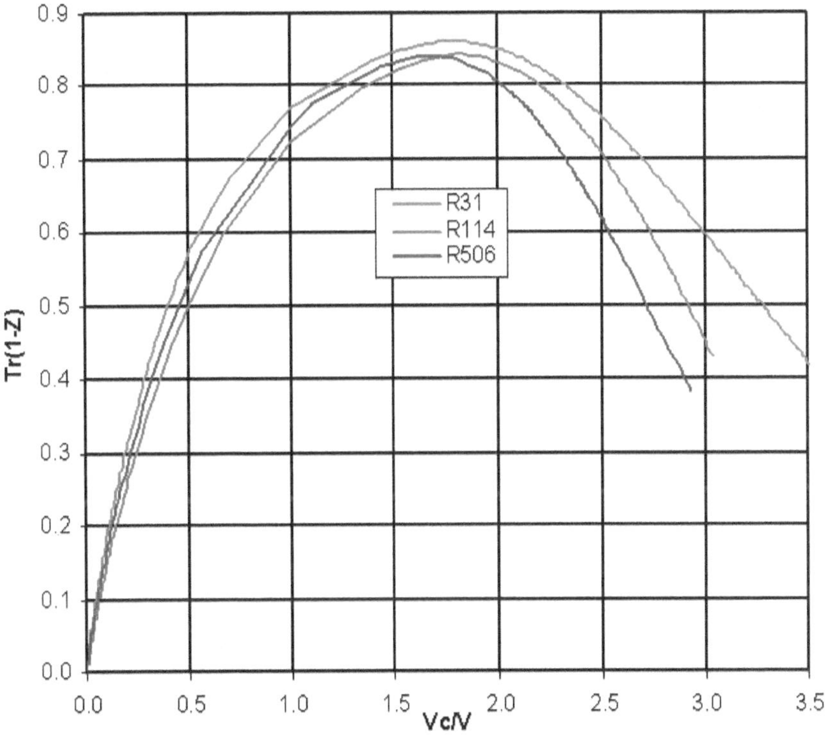

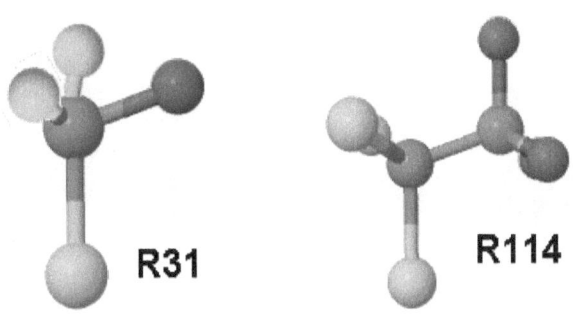

The molecules are depicted in the following figure, although these are not to scale.

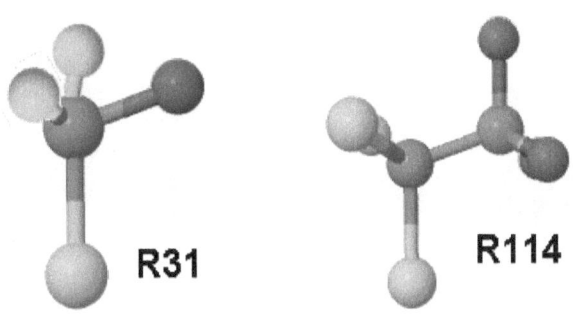

R31   R114

Chapter 3. Vapor Compression Refrigeration Cycle

The first working vapor-compression refrigeration system was built in 1834. The design came from Oliver Evans.[4] The first working model was built by John Gorrie.[5] Twenty years later, the first commercial ice-making machine was invented. Refrigerators for home use came out in 1923, when Frigidaire introduced the first self-contained unit. These used ammonia as a working fluid, which was dangerous. Many injuries and deaths motivated the search for safer alternative fluids. Freon (trademarked by the Chemours Company, a spin-off of Dupont) began production in the 1920s. This safer alternative rapidly caught on, especially for non-commercial applications, although industrial processes and ice skating rinks continued to use ammonia into the late twentieth century. The equipment is illustrated schematically in the following figure:

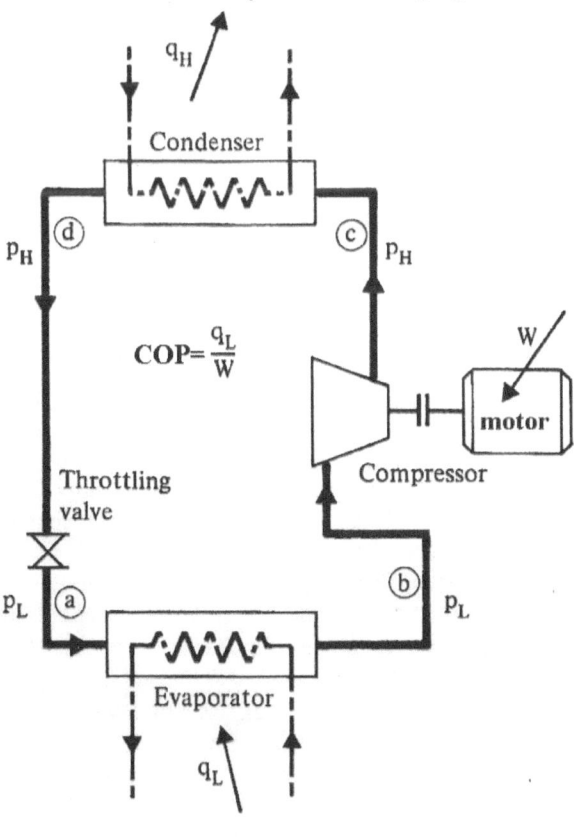

---

[4] Oliver Evans (1755-1819) American inventor who pioneered the high-pressure steam engine and created the first continuous production line.

[5] John B. Gorrie (1803–1855) American physician, scientist, inventor of mechanical cooling, and humanitarian.

The processes are often illustrated on a T-S (temperature vs. entropy) diagram, as illustrated below:

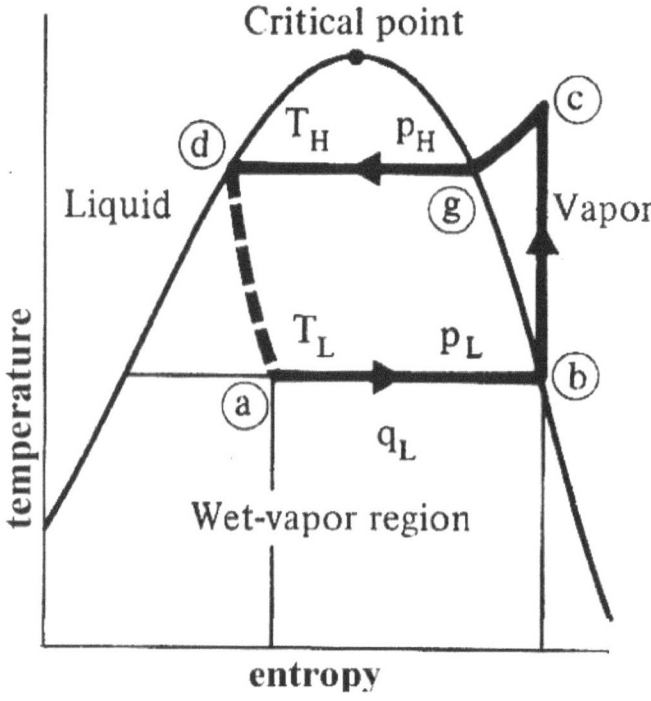

The points (**a**, **b**, **c**, **d**) on these two figures correspond. The process from **a** to **b** occurs in the evaporator, where heat, q_L, is removed from the low temperature reservoir (what you're trying to chill; located under the dashboard in a car air conditioning system). The process from **b** to **c** occurs in the compressor, requiring work, **w**. The process from **c** to **d** occurs in the condenser (in front of the radiator in a car), where heat, q_H, is removed from the high temperature reservoir (typically dissipated into the environment except in the cases of recovery or multi-stage systems). The working fluid is *throttled* from **d** back to **a** in what is called the *expansion valve*.

From the second figure we see that the process **a** to **b** occurs under the vapor dome along a roughly isothermal (and thus isobaric or constant pressure) line, ending at the saturated vapor state. The process **b** to **c** occurs in the superheated vapor region. In a real system, this segment bends toward the right (increasing entropy), as no real compression is isentropic. The process **c** to **g** occurs along an isobar (line of constant pressure) until the saturated vapor state is reached, where it continues along the same isobar, which is now also an isotherm, because the fluid is under the vapor dome, where temperature and pressure are dependent (i.e., not independent). This process continues inside the

18

condenser until reaching point **d**, the saturated liquid. The throttling process from **d** to **a** is isenthalpic (i.e., constant enthalpy), so it isn't isothermal or isentropic and is not a straight line on this chart. This is a pressure drop, so it isn't isobaric either.

Notice in the preceding figure that most of this cycle occurs under the vapor dome (the curve with the critical point at the top). Note also that $H_{FG}=T_{SAT}S_{FG}$, that is, the latent enthalpy (heat) of vaporization is equal to the saturation temperature times the latent entropy of vaporization. The width of the horizontal lines under the vapor dome along the X-axis is equal to S_{FG} and proportional to H_{FG}. This means that the area of the polygon **abgd** is proportional to the work and cooling load per unit mass flow rate; thus, we want a fluid with a wide vapor dome and saturation temperatures that bracket our low and high temperature reservoirs (the thing we're trying to cool and the sink to which we are dissipating the heat) in order to develop an effective system. Our choice of working fluids is not blind or random, but directed with purpose.

Joule-Thompson Effect

While you could achieve cooling entirely with a vapor and even with a liquid, this would be extremely ineffective and inefficient, especially the liquid. Humans can blow warm or cold air through their lips, so we inherently understand how refrigeration works. A pressure drop (through tight lips) produces a drop in temperature in many fluids, including air. The pertinent factor is that the same drop in pressure yields a range of temperature changes from negative to positive, depending on the fluid and conditions. This change in temperature is called the Joule-Thompson Effect, named after James Joule[6] and William Thompson.[7] The expansion valve is essentially isenthalpic (constant enthalpy; very little heat transfer) so that we have the Joule-Thompson coefficient, which characterizes this process so crucial to refrigeration:

$$\mu_{JT} = \left(\frac{\partial T}{\partial P} \right)_{H} \tag{3.1}$$

Equation 3.1 is the partial of temperature with respect to pressure at constant enthalpy. For positive values, if the pressure drops, so does temperature. In order to be effective (and efficient), we seek fluids with large positive values—the larger the better. Considering the process occurs entirely

---

[6] James Prescott Joule (1818–1889) English physicist, mathematician and brewer; studied the nature of heat and discovered its relationship to mechanical work, leading to the Law of Conservation of Energy.

[7] William Thomson, a.k.a. Baron Kelvin (1824–1907) Irish-Scottish mathematical physicist and engineer; instrumental in discovering and elucidating the First and Second Laws of Thermodynamics.

under the vapor dome, we also seek fluids with large latent heat. Ammonia meets both of these requirements quite well, better than most other refrigerants.

Compressor Power

The power associated with a flowing process is given by the following expression:

$$\frac{\dot{W}}{\dot{m}} = w = \int V dP \tag{3.2}$$

The pressure difference ($\int dP$) is given, arising from the difference in saturation pressures corresponding to the temperatures. The power is proportional to the specific volume. Most engineers and scientists don't realize that in a typical steam power plant having three sections of steam turbine, the low-pressure section produces more power than the high or intermediate pressure sections, typically 45% of the total. This is in spite of the fact that the pressure drops from about 200 atm to about 30 atm in the high pressure section, down to about 5 atm through the intermediate pressure section, and finally to about 1/30th atm at the exhaust where steam enters the condenser.

When we are extracting power, it is advantageous to have a large specific volume. The opposite is true when compressing. We don't want an extremely large specific volume entering the compressor of a refrigeration cycle, as this would mean more power consumed. Many people also don't realize that the molecular weight of water is 18, while that of air is 29 (think: ideal gas density $\rho = P/RT$ and the gas constant, R, is equal to a constant divided by the molecular weight). This is why water vapor is lighter than air and clouds float. We don't want to select a refrigerant whose saturation pressure corresponding to our low temperature reservoir is very small, as this would mean very large specific volume and very large compressor power. This consideration must direct our choice of fluids.

Comparison of Fluids

With thermodynamic properties in hand, we could identify the most advantageous temperature range for each one and also the unit work and heat (refrigeration load). This evaluation has been performed many years ago by several investigators for most common fluids, even some uncommon or unlikely ones. We can gather and augment this list so as to provide a comprehensive comparison of fluids. When designing a system, we first determine the target temperatures and the required heat load. We can then pick from one or more fluids that are suited to our needs. This is precisely why R12 was used in automotive applications and R22 was used in household refrigerators and freezers for decades. While these fluids have fallen out of use over concerns of global warming, this transition has not been driven by performance or efficiency. The following table is one such comparison of fluids, in this case an assortment that work well between a fixed pair of temperatures and a fixed cooling load of one ton.[8]

comparison of performance for 1 ton (3.52 kW) of refrigeration						
fluid		comp-ression ratio	specific cooling kJ/kg	mass flow kg/hr	comp. power kW	coef. of perf.
refrigerant number	chemical name					
R170	Ethane	2.86	136	92.8	1.456	2.41
R744	Carbon Dioxide	3.15	129	98.5	1.372	2.56
R744A	Nitrous Oxide	3.03	198	64.0	0.977	3.60
R115	Chloropentafluoromethane	3.89	68	187.2	0.872	4.02
R13B1	Bromotrifluoromethane	3.36	68	186.7	0.768	4.25
R600a	Isobutane	4.54	259	48.7	0.808	4.36
R502	22/115 azeotrope	3.75	106	119.2	0.805	4.37
R114	Dichlorotetrafluoroethane	5.42	100	126.3	0.782	4.49
R1270	Propylene	3.51	402	29.9	0.78	4.51
R290	Propane	3.70	281	44.9	0.768	4.58
R500	12/152a azeotrope	4.12	141	89.8	0.753	4.65
R22	Chlorodifluoromethane	4.03	163	77.8	0.754	4.66
R12	Dichlorodifluoromethane	4.08	116	108.9	0.747	4.70
R717	Ammonia	4.94	1103	11.5	0.737	4.76
R630	Methylamine	6.13	707	18.0	0.729	4.81
R1120	Trichloroethylene	11.65	213	59.3	0.731	4.82
R1130	Dichloroethylene	8.42	266	47.6	0.726	4.83
R113	Trichlorofrifluoroethane	8.02	125	101.5	0.726	4.84
R764	Sulfur Dioxide	5.63	329	38.4	0.722	4.87
R40	Methyl Chloride	4.48	349	36.2	0.717	4.90
R30	Methylene Chloride	8.60	313	40.6	0.718	4.90
R600	Butane	5.07	299	42.5	0.711	4.95
R21	Dichlorofluoromethane	5.96	208	61.0	0.702	5.01
R11	Trichlorofluoromethane	6.19	155	81.4	0.699	5.03
R160	Ethyl Chloride	5.83	331	39.5	0.676	5.21
R631	Ethylamine	7.40	525	24.2	0.638	5.52
R611	Methyl Formate	7.74	440	28.8	0.633	5.65
R610	Ethyl Ether	8.20	294	43.0	0.613	5.74
based on Tc=250°K and Th=300°K						

---

[8] One ton of refrigeration is the cooling derived from one ton (2000 pounds) of ice melting over a 24-hour period and has been the standard of unit refrigeration since the invention of such systems, which were originally "ice houses" (businesses that sold ice). It is equal to 3.516852842 kW and is the same energy as would be required to freeze this same amount of water in the same period of time.

We see from this table that there is considerable variation in performance even for refrigerants suited to these temperatures. The compression ratio (compressor outlet over inlet pressure) varies by a factor of 4. The specific cooling varies by a factor of 16. The power consumption varies by a factor of 2.4 and the coefficient of performance also varies by a factor of 2.4. Keep this in mind when considering swapping out a current refrigerant with one that meets some other objective than performance, such as reduced global warming potential.

The cycle tab in spreadsheet examples.xls (in the examples folder of the online archive) allows you to select a refrigerant from the database, set the operating conditions, and see the results, numerically and graphically. The calculation area is shown below:

	A	B	C	D	E	F	G
1	**Vapor-Compression Refrigeration Cycle**						
2	User Inputs		R1270	Refrigerant Number			
3	Units		4	SI kPa			
4	EvaporatorTemp.		-146.1	°C			
5	CondenserTemp.		33.1	°C			
6	Cooling Load		3.52	kW			
7	compressor eff.		70%	-			
8	motor efficiency		50%	-			
9	Calculations	T	P	V	h	s	x
10	point	°C	kPa	m³/kg	kJ/kg	kJ/kg/°C	quality
11	a	-146	0	2.E-03	413.4	3.2562	75%
12	b	-146	0	1531	550.1	4.3321	100%
13	c	404	1403	0.1	1587.3	4.8362	***
14	g	33	1403	0.033	729.9	3.0483	100%
15	d	33	1403	2.E-03	413.4	2.0146	0%
16	a	-146	0	0.0	413.4	3.2562	75%
17	b	-146	0	1531.4	550.1	4.3321	100%
18	c'	286	1403	0.077	1276.1	4.3321	***
19	flow	0.0257	kg/s				
20	compressor	26.69	kW				
21	electrical	53.38	kW				
22	coef.of perf.	0.066	-				
23	area	587.0	kJ/kg				

As you change the user inputs (bold blue text), the cycle is updated numerically and graphically, as shown in the following figures.

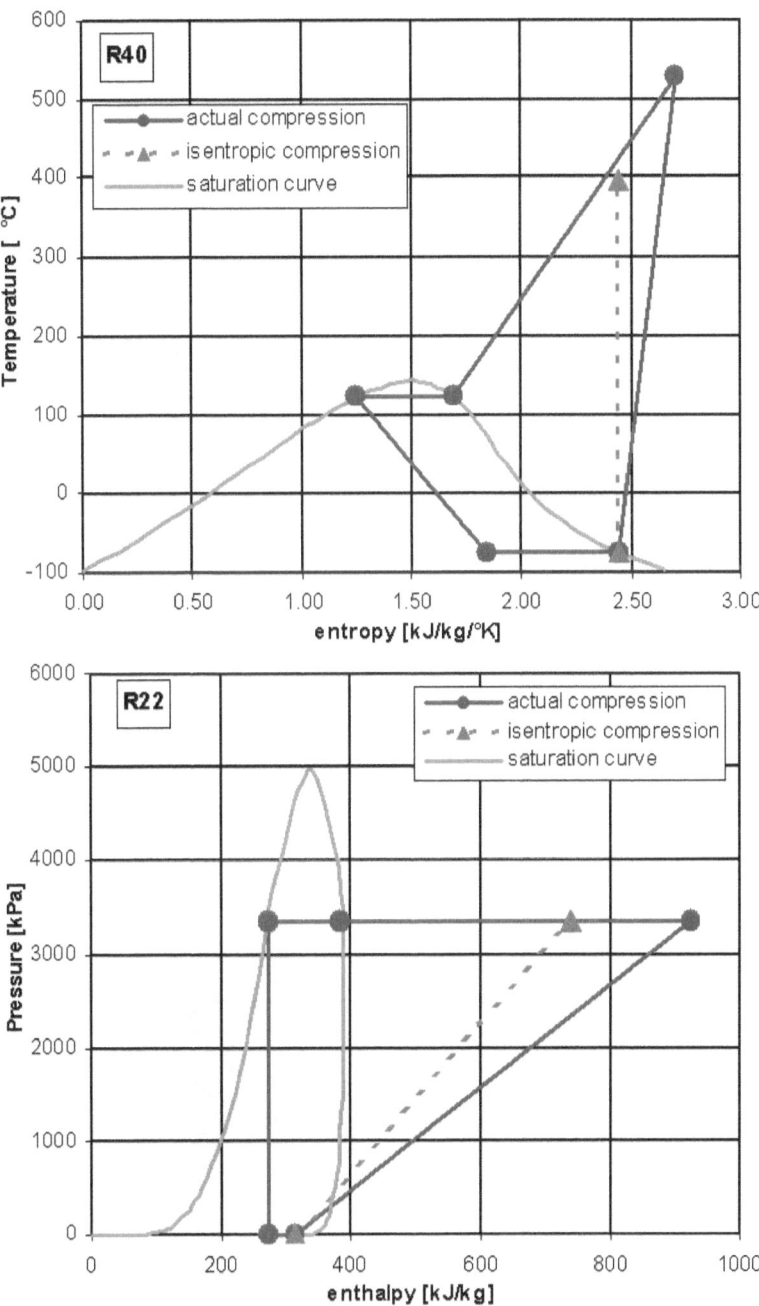

Chapter 4. Fluid Selection

We saw in Chapters 1 and 2 that some fluid behavior is more interesting than others, especially azeotropes. We might be able to take advantage of this interesting behavior. We saw in Chapter 3 that the selection of working fluid impacts the performance of the system. We now have a scientific way of determining the optimal conditions for each fluid when used in a vapor compression refrigeration system. This provides us with the optimum fluid for a given application—based on science. Of course, this doesn't consider non-scientific (i.e., political) motivations. Politicians are not known for their scientific acumen or perspicacity.

Available Functions in Refrigerant Excel AddIn

function	purpose
RefgName(index)	name
RefgLongName(index)	long name
RefgForm(index)	formula
RefgMW(name)	molecular weight (specify name or formula)
RefgZc(name)	critical compressibility
RefgTmax(name,units)	maximum temperature (for equation of state)
RefgPmax(name,units)	maximum pressure (for equation of state)
RefgPsat(name,Tsat,units)	saturation pressure (specify large positive number for critical value)
RefgTsat(name,Psat,units)	saturation temperature (specify large negative number for freezing point)
RefgVf(name,Tsat,units)	specific volume of saturated liquid (specify large positive number for critical value)
RefgVg(name,Tsat,units)	specific volume of saturated vapor (specify large positive number for critical value)
RefgPofTV(name,T,V,units)	pressure from temperature and specific volume
RefgVofTP(name,P,T,units)	specific volume from temperature and pressure
RefgHf(name,Tsat,units)	enthalpy of saturated liquid (specify large negative number for zero reference)
RefgHg(name,Tsat,units)	enthalpy of saturated vapor (specify large positive number for critical value)
RefgHofTV(name,T,V,units)	enthalpy from temperature and specific volume
RefgHofTP(name,P,T,units)	enthalpy from temperature and pressure
RefgMuofTP(name,P,T,units)	dynamic viscosity (m) from temperature and pressure
RefgNuofTP(name,P,T,units)	kinematic viscosity (n=m/r) from temperature and pressure
RefgSf(name,Tsat,units)	entropy of saturated liquid (specify large negative number for zero reference)
RefgSg(name,Tsat,units)	entropy of saturated vapor (specify large positive number for critical value)
RefgSofTV(name,T,V,units)	entrop from temperature and specific volume
RefgSofTP(name,P,T,units)	entropy from temperature and pressure
RefgSt(name,Tsat,units)	surface tension (vapor against liquid - only valid at saturation)
RefgTk(name,Tsat,units)	thermal conductivity
RefgTofPH(name,P,H,units)	temperature from pressure and enthalpy
RefgTofPS(name,P,S,units)	temperature from pressure and entropy
RefgZofTP(name,T,V)	compressibility from temperature and pressure
RefgZofTV(name,T,V)	compressibility from temperature and specific volume

We must stay well below the critical point, otherwise the cycle polygon (on a T-S diagram) is too small and the required flow rate would be too high. We must also stay well above the freezing point, as we don't want the working fluid to solidify. We have thermodynamic properties for dozens of refrigerants (see spreadsheets in the online archive). We have an Excel AddIn with various function calls to access these properties (see Refrig100 in the online archive).

We can use the Excel Solver to optimize the conditions for any of these fluids to minimize the specific power consumption or maximize the coefficient of performance or minimize the flow rate or some combination of these parameters. The problem with all of these is that they lead to trivial or undesirable results. The minimum power and flow rate will occur with any trivial temperature difference, which is pointless for refrigeration. The maximum coefficient of performance will also occur at useless combinations of high and low temperatures. What then is the appropriate criterion for optimization? The area of the cycle polygon on a T-S diagram. Think: Carnot efficiency and net work. The most advantageous operating conditions for a particular fluid are those that correspond to the largest polygon, which has units of energy per mass or the same as enthalpy.

$$area = \oint T dS \qquad (4.1)$$

We first run a series of cases sweeping through the available temperatures from the freezing point to the critical point, keeping the optimal solution (largest area). After the sweep is over, the best solution is copied into the spreadsheet. The resulting table is on the next page. With this table we can select which fluid would be optimal from a purely thermodynamic point of view. There are other considerations, including toxicity, flammability, lubrication, expense, and, of course, global warming potential. Still, our choice should not be random or made of convenience, as if this information weren't available.

The AddIn uses the modified Benedict-Webb-Rubin 32-parameter EoS described in Appendix D. The refrigerants are shown on a temperature-temperature scale on the page after the table. Notice that these lie along a roughly diagonal line. Azeotropes are shown in red and the rest in blue. Notice that the azeotropes are all clustered in one spot. Notice also that there are blue non-azeotropes near each of the red azeotropes, which means that from purely thermodynamic considerations, there is no compelling reason to select an azeotrope over a non-azeotrope. Other considerations, such as reduced global warming potential, most often influence the decision to use an azeotrope.

26

Table 4.1 Comparison of Refrigerants

refrigerant	Tl	Th	flow	power	COP	area
number	°C	°C	kg/s	kW	-	kJ/kg
R11	-105.5	75.0	0.049	28.71	0.12	157.2
R12	-152.1	39.0	0.075	71.34	0.05	282.3
R13	-176.1	7.4	0.179	200.86	0.02	362.4
R13b1	-162.8	18.0	0.145	117.60	0.03	244.7
R14	-146.1	-53.6	0.082	21.51	0.16	68.2
R21	-68.2	109.3	0.039	22.07	0.16	146.3
R22	-152.4	42.4	0.043	71.39	0.05	589.8
R23	-152.4	42.4	0.043	71.39	0.05	589.8
R31	-152.8	33.4	0.017	35.42	0.10	755.7
R40	-92.7	123.9	0.033	54.12	0.06	475.6
R50	-177.5	-87.6	0.023	33.77	0.10	472.8
R113	-16.4	109.7	0.079	17.31	0.20	34.9
R114	12.9	85.6	0.062	7.42	0.47	18.5
R115	-94.4	10.0	0.076	13.36	0.26	38.0
R134a	-98.3	37.5	0.041	18.69	0.19	110.9
R142b	-125.4	35.2	0.038	28.78	0.12	210.9
R152a	-113.6	54.3	0.028	30.04	0.12	288.4
R170	-177.8	18.1	0.056	226.77	0.02	1251.8
R290	-182.6	27.6	0.036	162.65	0.02	1429.2
R500	-140.0	25.5	0.041	36.54	0.10	331.4
R502	-101.1	20.2	0.049	15.11	0.23	77.1
R503	-146.1	-3.5	0.069	43.22	0.08	174.5
R504	-138.6	34.5	0.053	46.26	0.08	249.3
R505	-118.2	45.1	0.047	26.00	0.14	153.0
R506	-109.4	69.9	0.039	33.51	0.10	237.5
R600	-133.3	37.0	0.029	39.78	0.09	336.6
R600a	-154.4	17.2	0.027	48.26	0.07	498.0
R702	-254.2	-245.0	0.010	7.62	0.46	273.4
R704	-270.7	-268.2	0.310	24.09	0.15	26.1
R717	-72.7	90.0	0.005	16.43	0.21	1011.0
R718	5.0	369.1	0.006	65.15	0.05	3939.7
R720	-243.6	-233.7	0.060	5.56	0.63	33.9
R728	-205.0	-152.0	0.043	22.32	0.16	193.0
R729	-193.3	-145.5	0.033	13.20	0.27	113.8
R732	-213.8	-127.0	0.048	50.33	0.07	409.9
R740	-184.3	-127.5	0.045	17.83	0.20	166.4
R744	-51.6	26.0	0.023	7.60	0.46	96.0
R1150	-146.1	4.2	0.032	59.78	0.06	550.9
R1270	-146.1	33.1	0.026	53.38	0.07	587.0

Refrigerant Optimal Temperature Zones

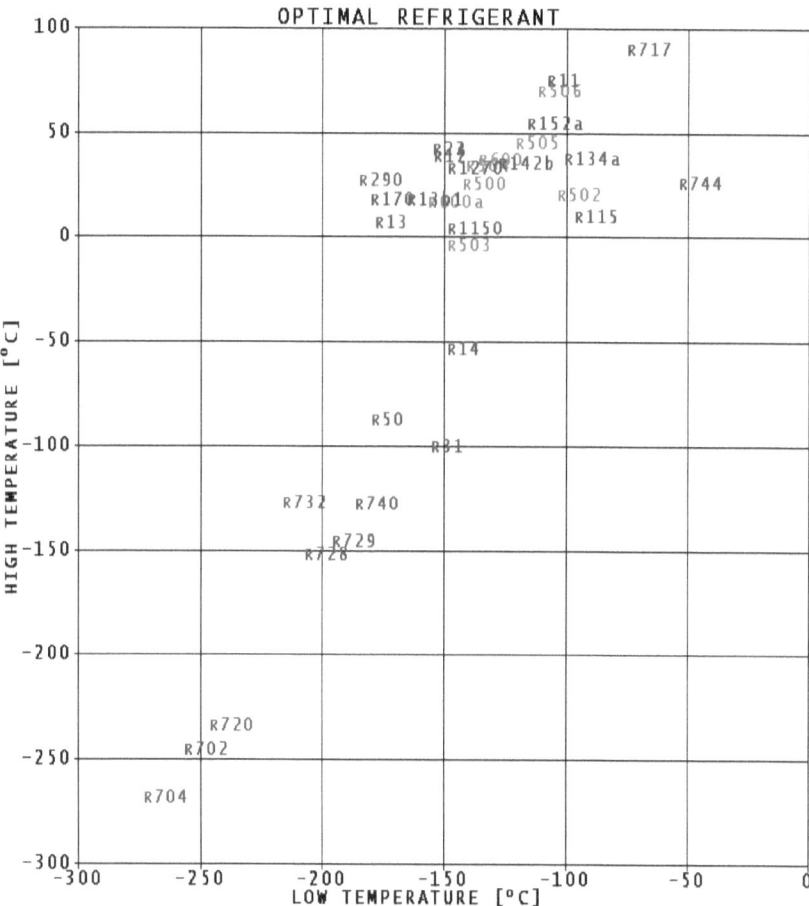

Chapter 5. Vapor Expansion Power Cycle

While refrigerants are most often used in vapor compression cycles to provide cooling, use in vapor expansion cycles to produce power is gaining interest, especially in heat recovery systems and geothermal as well as some concentrated solar applications, although interest in these last is waning. A simple Rankine cycle is depicted in the following figure. The components are labeled along the process segments comprising the cycle.

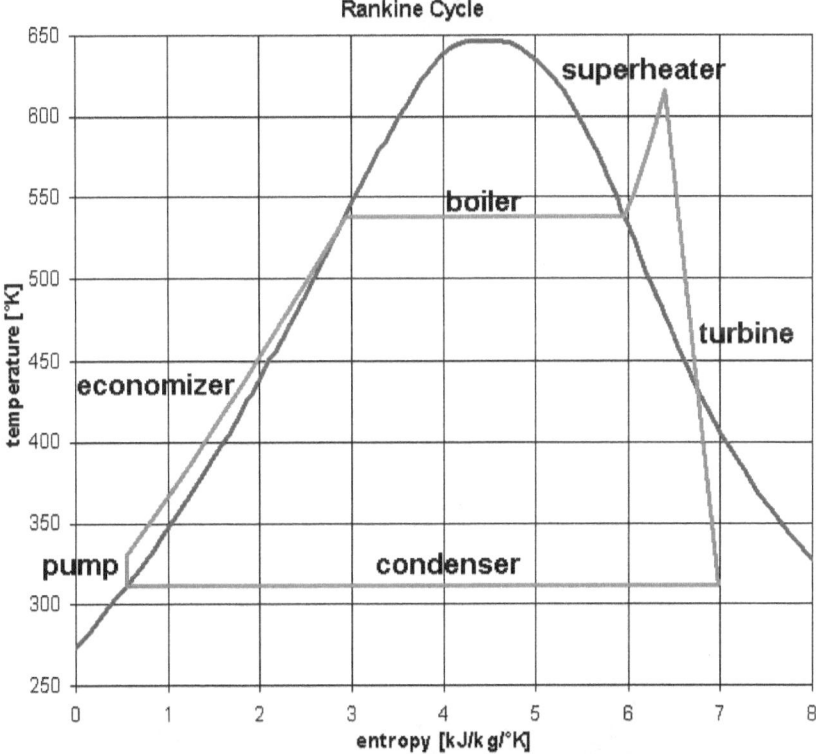

The biggest difference in the shape of the polygon between the vapor expansion power cycle and the vapor compression refrigeration cycle is on the bottom left side. The liquid is always fully condensed before entering the pump, not only to avoid cavitation but also to minimize the power consumption (Equation 3.2 is applicable here). The economizer process roughly follows an isobar, although there is always some pressure drop. The superheater line is also roughly isobaric with some pressure drop. All real pumps and turbines have an efficiency less than 100% and so generate entropy, which means that these process lines curve to the right.

The spreadsheet for power is set up similarly to the one for refrigeration and is on the Rankine tab. The user inputs are again in bold blue type plus there are the two buttons to run all cases for one fluid and run all fluids. The calculation area is shown below:

	A	B	C	D	E	F	G
1	**Vapor-Expansion Power (simple Rankine) Cycle**						
2	user inputs		R12	Refrigerant Number			
3	units		4	SI kPa			
4	condenserTemp.		-132.6	°C			
5	evaporatorTemp.		89.7	°C			
6	superheat		0.0	°C			
7	desired power		3.52	kW			
8	pump efficiency		65%	-			
9	turbine efficiency		85%	-			
10	Calculations	T	P	V	h	s	x
11	point	°C	kPa	m³/kg	kJ/kg	kJ/kg/°C	quality
12	a	-133	2.4.E-02	5.7.E-04	20.7	0.1623	0%
13	b	-133	2.4.E-02	5.7.E-04	23.2	0.1623	***
14	c	90	2.8.E+03	1.0.E-03	230.4	1.0210	0%
15	d	90	2.8.E+03	5.3.E-03	312.0	1.2457	100%
16	e	90	2.8.E+03	5.3.E-03	312.0	1.2457	***
17	f	-133	2.4.E-02	5.7.E-04	193.8	1.3941	84%
18	f'	-133	2.4.E-02	5.7.E-04	173.0	1.2457	74%
19	b'	-132	2.8.E+03	5.7.E-04	22.3	0.1623	***
20	flow	0.0304	kg/s				
21	turbine	118.1	kJ/kg				
22	pump	-2.4	kJ/kg				
23	net power	115.7	kJ/kg				
24	heat in	288.8	kJ/kg				
25	heat out	-173.1	kJ/kg				
26	efficiency	40.1%					
27	area	161.9	kJ/kg				

There are four additional lines for the pump and net work as well as both heat transfers (in and out); whereas, before we didn't have a separate line for the expansion valve. The turbine exit quality (1-moisture) is in cell G17. The line f' is the isentropic turbine exit conditions and b' is the isentropic pump exit conditions, which are calculations and not actual state points in a real system.

The temperature-entropy diagram updates automatically when you change the fluid, units, or any of the other user inputs:

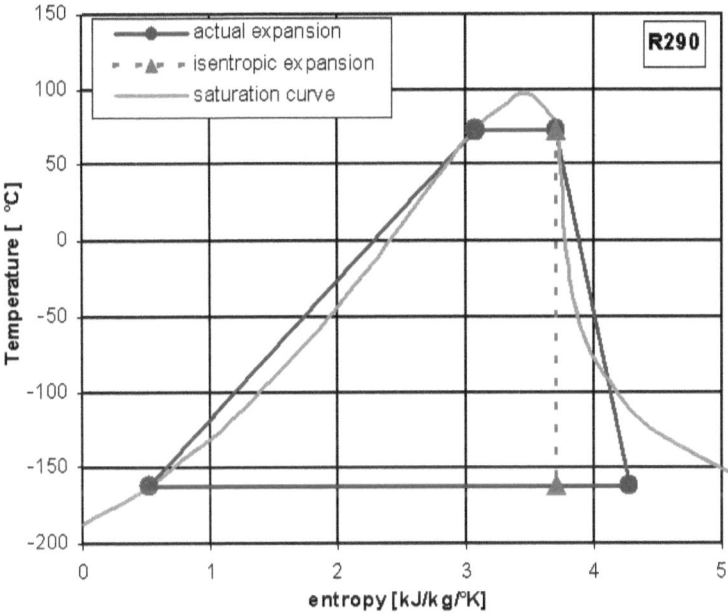

The pressure-enthalpy curve also updates automatically:

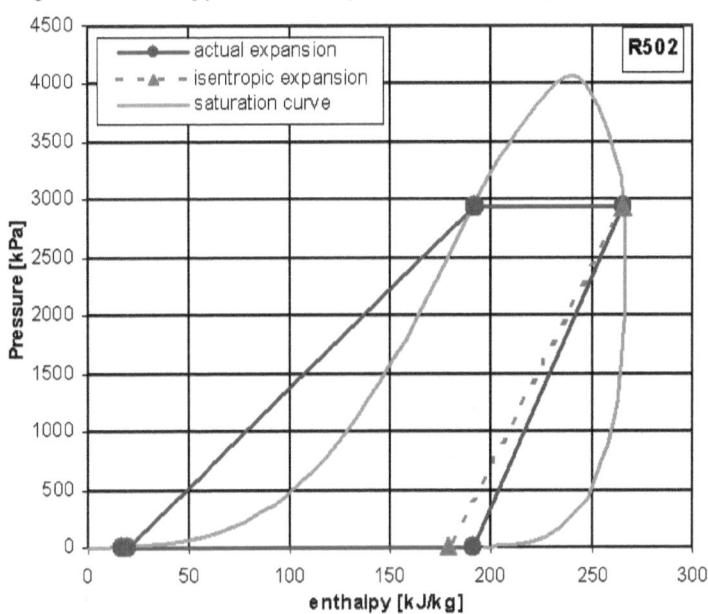

If the polygons were the same, then the optimal temperatures for use in a power cycle would be the same as for refrigeration as far as the fluid choice was concerned. Because the polygons are not the same shape, we get slightly different results. These are generated in the same way in the spreadsheet (examples.xls) with Excel macros. The degree of superheat represents a complication and an additional variable. Turbines don't like moisture (condensation). This greatly decreases their lifetime (see blade erosion in figure below). We superheat the vapor to avoid dropping too far into the wet region. We could add maximum allowable moisture as a constraint, to control the required superheat. The lowest quality (highest moisture) is 70% for R720. This is a little too low to be practical. A more reasonable value would be at least 80%. It is not surprising that water (R718) has the largest area.

Comparison of Performance for 3.517 kW Net Output

refrigerant number	Tl °C	Th °C	moist %	flow kg/s	net kJ/kg	eff kJ/kg	area kJ/kg
R11	-99.5	172.8	91%	0.026	132.94	0.39	186.6
R12	-135.9	89.7	83%	0.03	118.82	0.41	167.4
R13	-166.1	11.5	76%	0.035	99.67	0.41	140.3
R13b1	-146.6	51.2	81%	0.044	79.62	0.37	120.1
R14	-146.1	-59.2	81%	0.083	42.49	0.27	53.7
R21	-51.6	157.5	87%	0.034	103.79	0.31	135.3
R22	-148.5	72.6	72%	0.022	158.88	0.43	228.1
R23	-138.6	11	94%	0.032	110.81	0.1	551.1
R31	-152.8	145.9	83%	0.014	246.8	0.25	909.8
R40	-92.7	108.3	74%	0.018	193.34	0.36	251.5
R50	-177.5	-96.4	76%	0.019	183.82	0.31	239.4
R113	-13.6	195	123%	0.048	73.15	0.27	94.7
R114	8.8	133.7	118%	0.1	35.11	0.18	43.6
R115	-94.4	67.7	107%	0.055	63.81	0.29	82.8
R134a	-98.2	84	90%	0.031	112.05	0.32	147.7
R142b	-106.3	117.5	91%	0.024	144.91	0.37	196.8
R152a	-108.5	92	82%	0.019	180.55	0.36	242.0
R170	-174.6	14.4	70%	0.011	326.91	0.44	470.4
R290	-173.8	73.2	74%	0.009	388.46	0.49	586.4
R500	-131.7	81.7	80%	0.027	128.39	0.4	174.6
R502	-93.4	66.2	92%	0.048	73.97	0.3	95.5
R503	-146.1	5.4	77%	0.04	88.36	0.35	118.9
R504	-138.6	28.2	78%	0.028	126.79	0.39	161.3
R505	-106.6	95.7	86%	0.034	101.98	0.37	132.1
R506	-100.5	120.4	84%	0.025	140.77	0.36	193.7
R600	-133.3	138.3	95%	0.01	352.14	0.43	504.8
R600a	-148.6	110.8	90%	0.01	348.98	0.45	503.2
R702	-252.8	-247.8	86%	0.05	70.87	0.16	87.1
R704	-270.7	-268.5	73%	0.385	9.14	0.35	12.2
R717	-62.4	97.6	74%	0.008	442.87	0.29	566.6
R718	31.5	316.2	71%	0.004	858.16	0.33	1114.2
R720	-246.8	-232	70%	0.172	20.42	0.25	25.8
R728	-205	-156.1	75%	0.057	61.94	0.28	79.8
R729	-192.7	-152.2	71%	0.077	45.85	0.22	59.2
R732	-209.7	-135.1	71%	0.035	99.64	0.37	133.6
R740	-183.3	-133.2	72%	0.094	37.38	0.25	47.6
R744	-48.6	16.5	79%	0.076	46.28	0.15	57.9
R1150	-143.9	-16.9	75%	0.017	203.16	0.33	267.5

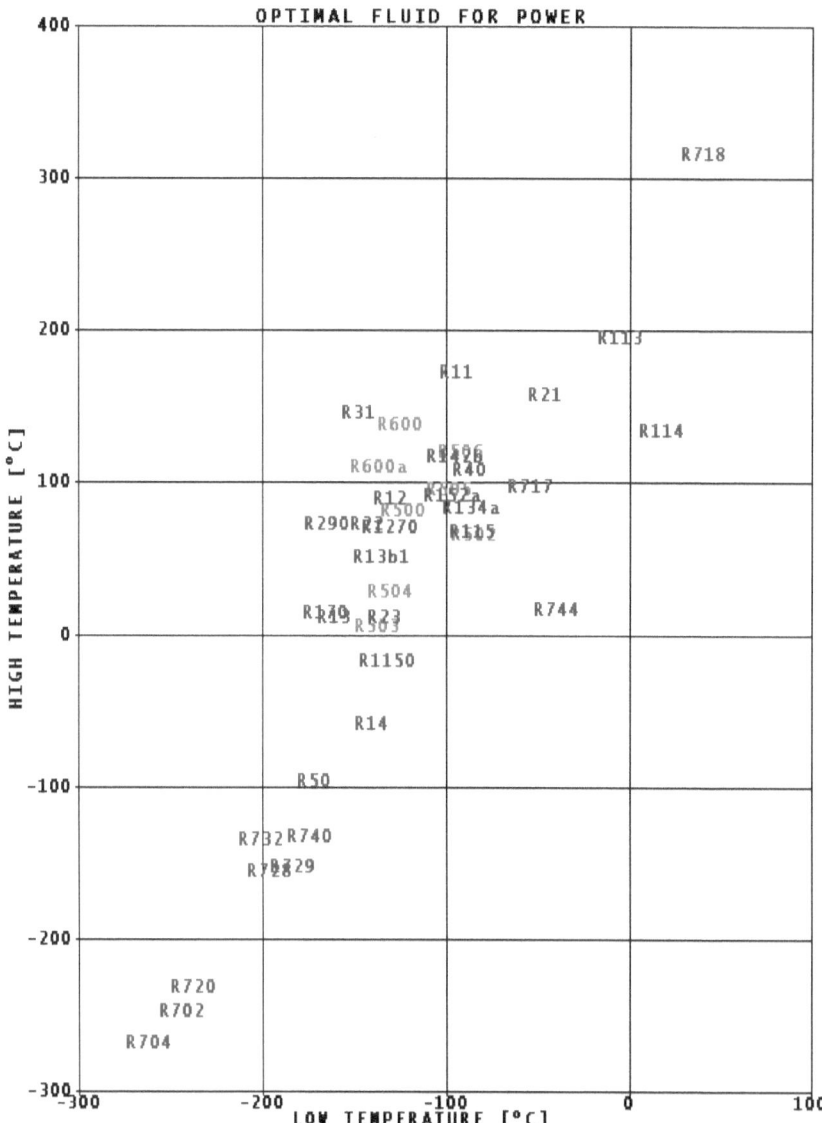

OPTIMAL FLUID FOR POWER

34

Chapter 6. Boyle Temperature and Dipole Moment

The Boyle temperature (named after Robert Boyle[9]) is where a real fluid behaves like an ideal gas at non-trivial pressure. It is best illustrated on a plot of compressibility ($Z=PV/RT$) vs reduced pressure, $P_R=P/P_C$. For an ideal gas, $Z=1$. The only nearly horizontal line is shown in magenta and corresponds to a reduced temperature, $T_R=T/T_C$, of approximately 2.5.

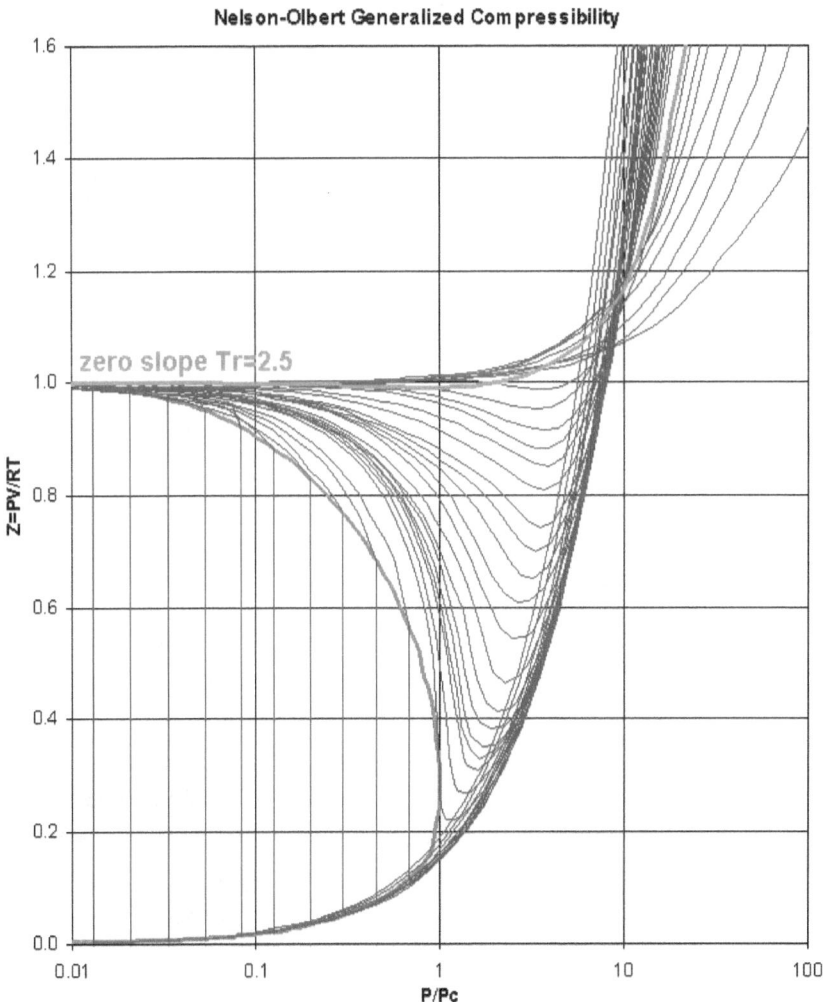

Nelson-Olbert Generalized Compressibility

---

[9] Robert Boyle (1627–1691) Anglo-Irish natural philosopher, chemist, physicist, and inventor, famous for his work with gases.

The Boyle temperature is often defined as the point at which the second virial coefficient is zero, but this is a little vague. It is also the point at (or zone over) which the attractive force (think: parameter a from Chapter 2) equals the repulsive force (think: parameter b from Chapter 2) so that they cancel each other out and the non-ideal behavior vanishes. The virial expansion is:

$$Z = \frac{PV}{RT} = 1 + \frac{B(T)}{V_R} + \frac{C(T)}{V_R^2} + \ldots \qquad (6.1)$$

The second virial coefficient is B(T) and is presumed to be a function of temperature alone. Much study has been made of the second virial coefficient and the Boyle temperature. Tsonopoulos presented one of the first successful correlations for the second virial coefficients of both polar and non-polar fluids, based on the Pitzer-Curl correlation.[10] Tsonopoulos' correlation contains three terms: 1) spherical molecule term (zero acentric factor; noble gases), 2) a non-polar term, and 3) a polar term. In doing so, Tsonopoulos reported that the second virial coefficient was linked to the dipole moment. The dipole moment of a molecule arises from separated charges, their strengths, and the bond length; thus we see a possible connection to the molecular interactions, which we know to be special in the case of azeotropes.

Several quantities are called *dipole moment* in the literature, including: *Bond* dipole moment, a measure of polarity of a chemical bond. *Electric* (or *molecular*) dipole moment, a measure of the electrical polarity of a system of charges. *Electron* dipole moment, a measure of the charge distribution within an electron. *Magnetic* dipole moment, a measure of the magnetic polarity of a system of charges. *Transition* dipole moment is used in quantum mechanics. Here we are interested in the *bond dipole moment*. Debye is the unit of measure for the dipole moment, which is 3.33×10^{-30} m°K.

Khoshsima and Hosseini provide Boyle temperatures for an extensive list of fluids, which we will consider in light of this.[11] We are focusing particularly on refrigerants, but several other fluids have been added to the list to extend the range of values. The data can be found in the online archive in spreadsheet Boyle.xls. Only the most relevant columns are listed in the following table. The rest are in the spreadsheet. The data are listed in order of increasing dipole moment.

---

[10] Tsonopoulos, C., "An Empirical Correlation of Second Virial Coefficients," AICHE Journal, Vol. 20, pp. 263–272, 1974.

[11] Khoshsima, A. and Hosseini, A., "Prediction of the Boyle Temperature, Second Virial Coefficient and Zeno Line Using the Cubic and Volume-Translated Cubic Equations of State," Journal of Molecular Liquids, Vol. 242, pp. 625–639, 2017. [This excellent article is available on the web and includes many useful equations.]

Boyle Temperatures & Dipole Moments		refrig	Zc	Boyle	dipole
name	formula	number	-	ºK	Debye
Carbon Tetrafluoride	CF4	R14	0.279	540.5	0.00
Methane	CH4	R50	0.286	510.4	0.00
Ethane	C2H6	R170	0.280	757.5	0.00
Carbon Dioxide	CO2	R744	0.275	686.4	0.00
Ethylene	CH2=CH2	R1150	0.281	400.0	0.00
Carbon Tetrachloride	CCl4	-	0.2679	1297.4	0.00
n-Butane	C4H10	R600	0.274	976.5	0.05
Propane	C3H8	R290	0.276	880.0	0.08
Carbon Monoxide	CO	-	0.3213	304.1	0.11
1-Butene (butylene)	C4H8	-	0.2770	1008.3	0.34
Propylene	CH3CH=CH2	R1270	0.275	910.5	0.37
Trichloromonofuoromethane	CCl3F	R11	0.279	1091.0	0.45
Chlorotrifluoromethane	CClF3	R13	0.277	707.5	0.50
Dichlorodifluoromethane	CF2Cl2	R12	0.276	897.2	0.51
Bromotrifluoromethane	CBrF3	R13b1	0.280	850.4	0.65
Hydrogen Sulfide	H2S	-	0.2832	940.6	0.97
Hydrogen Chloride	HCl	-	0.2493	1005.6	1.08
Dichlorofluoromethane	CHCl2F	R21	0.270	1033.6	1.29
Ammonia	NH3	R717	0.254	996.1	1.30
Chlorodifluoromethane	CHClF2	R22	0.268	837.1	1.42
Sulfur Dioxide	SO2	-	0.2684	1142.7	1.63
Trifluoromethane	CHF3	R23	0.259	657.6	1.65
Ethanol	C2H5OH	-	0.2480	1283.9	1.69
Methanol	CH3OH	-	0.1900	1094.7	1.70
Chlorofluoromethane	CH2FCl	R31	0.230	1062.3	1.82
Water	H2O	R718	0.233	1325.3	1.84
Nitric Acid	HNO3	-	0.2311	1399.7	2.17
Aluminium Iodide	AlI3	-	0.1795	2071.0	2.48
Aluminium Bromide	AlBr3	-	0.1412	1522.0	5.14
Silver Chloride	AgCl	-	0.3000	6468.4	5.73
Methyl Chloride	CH3Cl	R40	0.275	989.9	8.54

We will first consider the correlation between Boyle temperature and dipole moment. The correlation is not as good as we might think from surveying the literature. The regression coefficient is $R^2=0.7407$ (somewhat poor; certainly not excellent), even after excluding two outliers (Aluminum Bromide and Methyl Chloride).

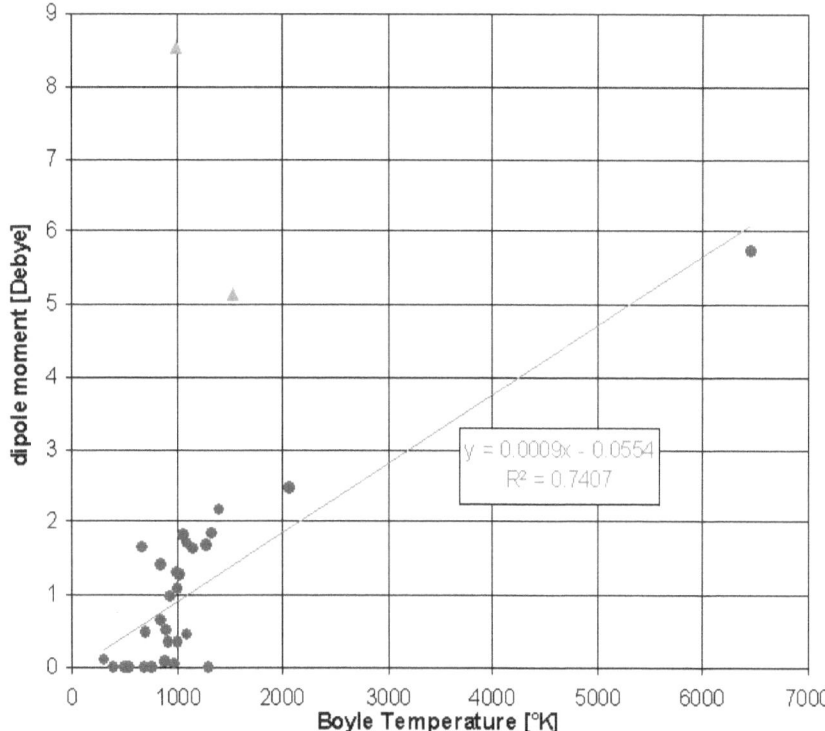

Several papers suggest the dipole moment might be more clearly related to the critical compressibility, Zc. This is shown in the figure below, also excluding two outliers (this time Silver Chloride and Methyl Chloride). The regression coefficient $R^2=0.7591$ is only slightly higher than for the previous.

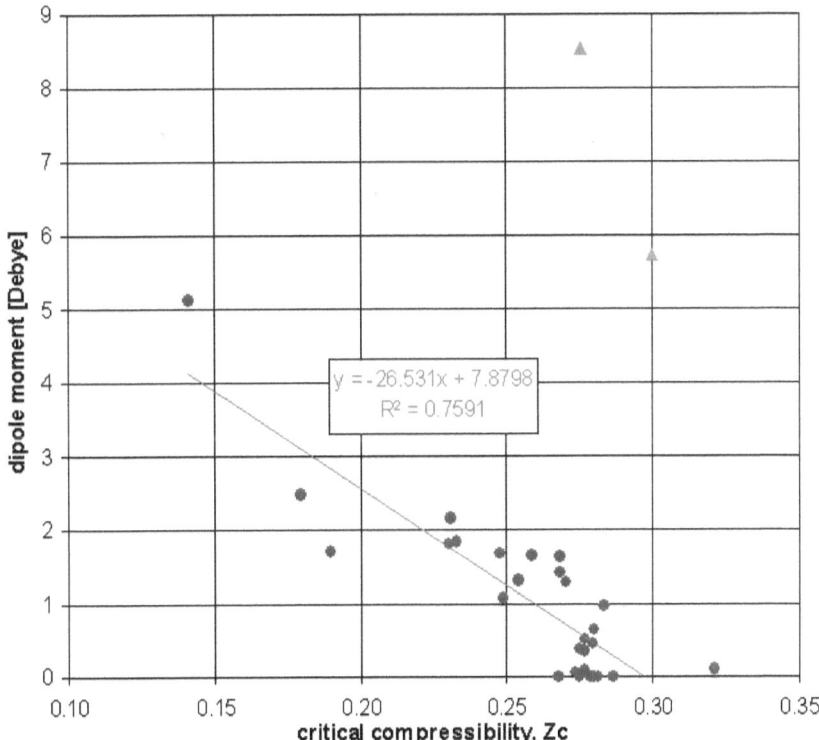

Before abandoning this approach, we consider a regression using both critical compressibility and Boyle temperature. The regression coefficient is $R^2=0.847726$, which is approaching an acceptable level. In this last case we exclude only one outlier (Methyl Chloride).

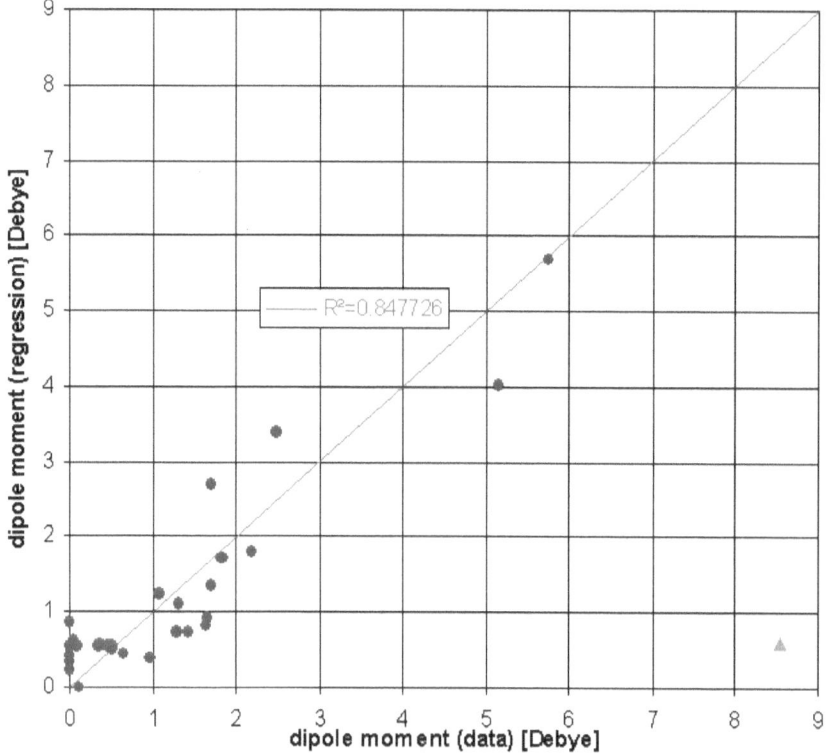

Of course, azeotropes aren't a single molecule so the traditional dipole moment doesn't apply. Still, we could calculate an effective dipole moment, except that none of these three correlations are sufficiently accurate to have much confidence in the result. Even with this uncertainty, we see no distinctive grouping of azeotropes when plotted against critical compressibility.

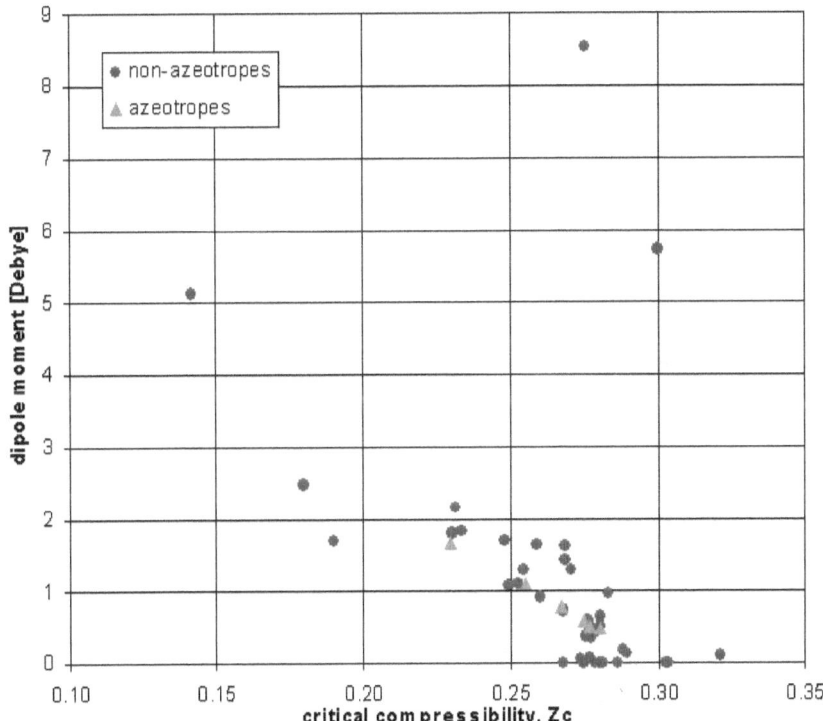

41

We also don't see any grouping of azeotropes when plotted against Boyle temperature.

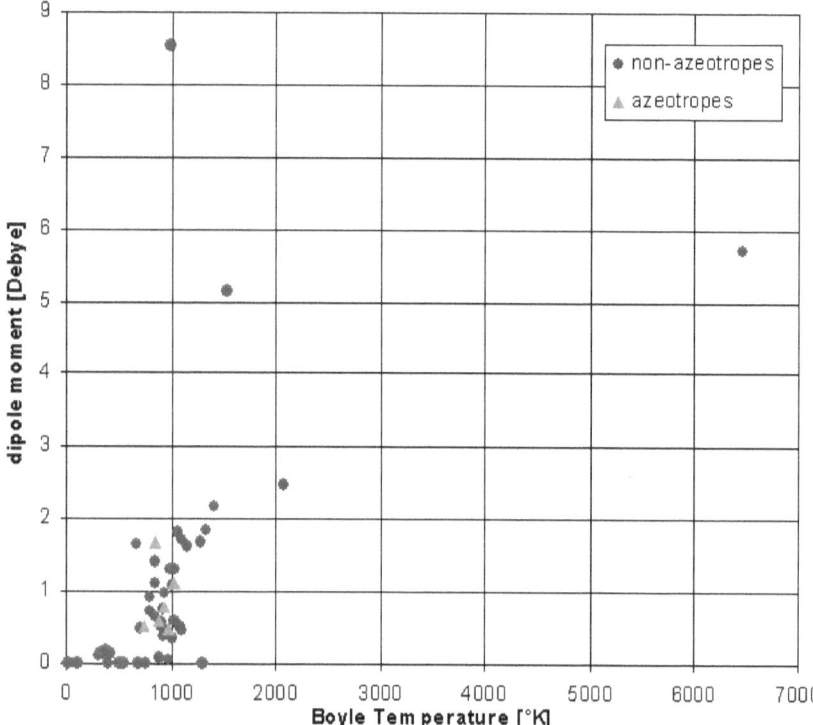

In summary, while we suspect there may be a connection between azeotropic behavior and dipole moment and there may be a connection with Boyle temperature and also possibly critical compressibility, these relationships are not sufficiently clear so that we might infer azeotropic behavior from them. So then it's back to the laboratory to try combinations and measure the properties until a better idea comes along.

Chapter 7. Lennard-Jones Potential

The Lennard-Jones potential (sometimes called LJ or 6-12) is a two-term approximation of the intermolecular potential proposed in 1924.[12] The two terms contain a sixth and twelfth power. The potential energy is given by:

$$\Gamma(r) = 4\varepsilon \left[\left(\frac{\sigma}{r} \right)^{12} - \left(\frac{\sigma}{r} \right)^6 \right] \tag{7.1}$$

where $\Gamma(r)$ is the potential, which varies with r, the radial distance between the particles, atoms, or molecules; ε is the strength; and σ is a length parameter, at which radius the two terms cancel.

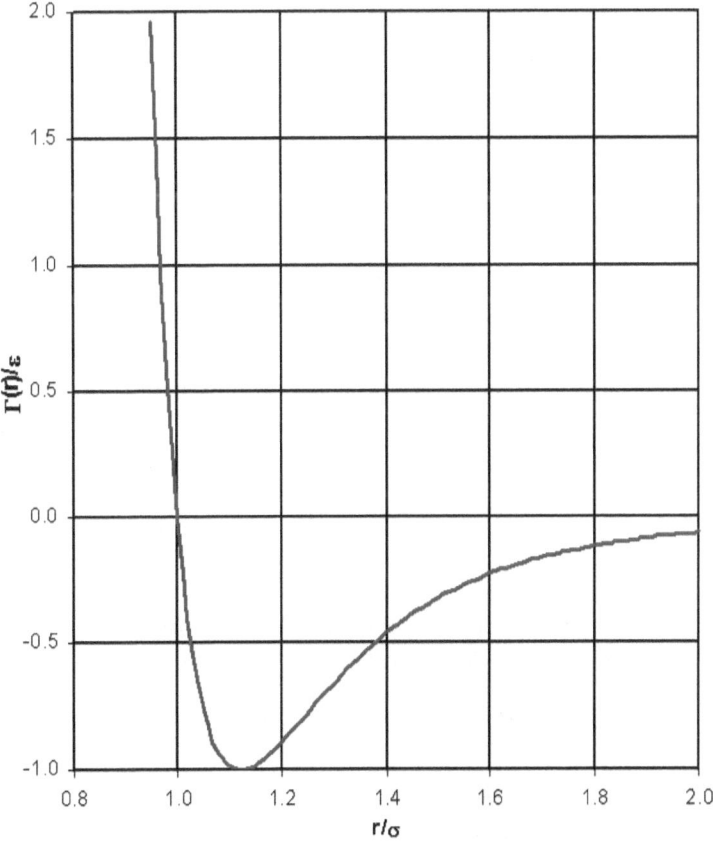

---

[12] John Edward Lennard-Jones (1894–1954) British mathematician and professor of theoretical physics at the Universities of Bristol and Cambridge, founder of modern computational chemistry.

The blue curve intersects $\Gamma(r)/\varepsilon=0$ at $r/\sigma=1$ and the minimum (equal to -1) occurs at $2^{1/6}$. We introduce this relationship here because it was the first entry in the field of computational chemistry and we hope through it to identify the anomalous behavior of azeotropes. We have reason to consider this because Prausnitz[13] has an entire section on mixing rules for the Lennard-Jones potential, including data and comparisons. Prauznitz also suggests that, based on fluid data, the first exponent should be somewhere between 9 and 12, so we have an additional parameter that can be tuned.

Two papers are most helpful here, one by Kolafa and Nezbeda[14] and a second by Thol, Rutkai, Köster, Lustig, Span, and Vrabec[15], both of which are available free online. Kolafa and Nezbeda (K&N) begin with the hard sphere model, which is a common introductory topic in statistical mechanics. The shape of the potential well is similar to LJ, only squared-off. This often begins with a definition of the packing fraction, η.

$$\eta = \frac{\pi \rho d^3}{6} \qquad (7.2)$$

where d is the hard sphere diameter and ρ is the density. The hard sphere compressibility is:

$$Z_{HS} = \frac{1+\eta+\eta^2-\frac{2}{3}\eta^3(1+\eta)}{(1-\eta)^3} \qquad (7.3)$$

As explained elsewhere (for example in either one of my books, *Thermodynamics* or *Thermodynamic and Transport Properties of Fluids*), all thermodynamic properties can be unambiguously and consistently derived from the Helmholtz Free Energy ($A=U-TS$). The HFE for the hard sphere potential is:

$$A_{HS} = T\left[\frac{5}{3}\ln(1-\eta)+\frac{\eta(34-33\eta+4\eta^2)}{6(1-\eta)^2}\right] \qquad (7.4)$$

The pressure and entropy are found by taking the partial derivative of the HFE with respect to density and temperature, respectively:

$$P = \rho T + \rho^2 \frac{\partial A}{\partial \rho} \quad S = -\frac{\partial A}{\partial T} \qquad (7.5)$$

[13] Prausnitz, J. M., *Molecular Thermodynamics of Fluid-Phase Equilibrium*, Prentice-Hall, 1969.
[14] Kolafa, J. and Nezbeda, I., "The Lennard-Jones Fluid: An Accurate Analytic and Theoretically-Based Equation of State," Fluid Phase Equilibria, Vol. 100, 1994.
[15] Thol, M., Rutkai, G., Köster, A., Lustig, R., Span, R., and Vrabec, J., "Equation of State for the Lennard-Jones Fluid," *Journal of Physical and Chemical Ref. Data*, 2016.

K&N then modify Equation 7.4 through computer simulation and regression to form an expression based on the LJ potential. They provide this EoS in the form of code (kolafanz.c), which you can find in the online archive in folder examples. The resulting PV isotherms are shown below:

The blue curves in the preceding figure are the van der Waals isotherms and the red curves are based on K&N's equations. The critical isotherm is noted above and the two critical isotherms are almost coincident. Near the end of this article (their Equation 34), we are told that the critical point of their EoS is at $T_C=1.3396$, $P_C=0.1405$, and $\rho_C=0.3108$, yielding $Z_C=0.3375$. The code (kolafanz.c) has been normalized and so have the isotherms in the preceding figure. While they (K&N) do not clearly explain this discrepancy, we can make some sense of it from the information provided. First, they refer to the applicability of their work as covering the range: $0.7<T<6$. Clearly, their symbol

T does not mean temperature (else we would be discussing liquid hydrogen), in spite of it being described as such in the List of Symbols on page 29 of their paper. It also doesn't mean $T_R=T/T_C$, otherwise the critical point would occur at unity. So what does it mean?

We next consider the paper by Thol, Rutkai, Köster, Lustig, Span, and Vrabec (TRKLSV). While this second paper perpetuates the same vague nomenclature and cites the first, it does provide the following definitions in Section 2.

$$T^* = \frac{kT}{\varepsilon}$$

$$p^* = \frac{p\sigma^3}{\varepsilon} \qquad (7.6)$$

$$\rho^* = \rho\sigma^3$$

where k is Boltzmann's constant (1.380649×10^{-23} J/°K). We are told in this same paragraph that $\rho=N/V$, where N is the number of particles and V is the volume; so this isn't density on a mass or molar basis. We know from Equation 7.1 that the length parameter, σ, has dimensions of length (on the order of a few Angstroms, 10^{-10} m) so that ρ^* is dimensionless, but not equal to $\rho_R=\rho/\rho_C$. This explains why the K&N ρ_C is equal to 0.3108 and the TRKLSV ρ_C is equal to 0.31.

The LJ parameter, ε, is approximately one kJ/mole. Substituting these quantities into Equation 7.6a, we find that $T^* \approx 10^{-25}$ mole°K. Perhaps these eight authors forgot to mention Avogadro's number, 6.022×10^{23} particles/gram-mole, which might put this parameter back into the range of unity. This parameter isn't dimensionless and certainly not equal to $T_R=T/T_C$, which explains why the K&N $T_C=1.3396$ and the TRKLSV $T_C=1.32$. Substituting these quantities into Equation 7.6b, we find that $p^* \approx 10^{-32}$ mole. Avogadro's number may be implied here too, so that the K&N $P_C=0.1405$ and the TRKLSV $P_C=0.13006$.

The TRKLSV paper presents a 90-parameter relationship between p^*, T^*, and ρ^* which is said to be an improvement over K&N. The shape is very similar to the preceding figure comparing the K&N isotherms to those of van der Waals. As these two results are a single relationship with fixed values of the various small-scale parameters, including a fixed exponent of 12, this exercise is of purely academic interest, for we have had numerous algebraic equations of state since van der Waals in 1873! We know that the van der Waals EoS doesn't fit any known fluid well, let alone a variety of fluids, especially azeotropes. This is quite discouraging, but...

Chapter 8. Prausnitz and the Virial Expansion

Three decades before Kolafa and Nezbeda and five decades before Thol, Rutkai, Köster, Lustig, Span, and Vrabec, Prausnitz successfully used the Lennard-Jones potential to explain the properties of mixtures, including the strange behavior of azeotropes. Prausnitz begins this discussion in Section 4.5 (ibid) with the potential and then introduces the Theory of Corresponding States (TCS) in Section 4.6. We have been implicitly using the TCS from the beginning of this text, by introducing the reduced temperature, pressure, and specific volume. In Section 4.7, Prausnitz notes that the TCS does not work for all substances (e.g., polar compounds and azeotropes), indicating that some additional parameter is needed. The Pitzer acentric factor and critical compressibility are the two most common. For the remainder of his Chapter 4, Prausnitz discusses exceptions and possible causes, including polarity and bonds.

In his Chapter 5, Prausnitz introduces the virial EoS and how the first and second virial coefficients are related to the intermolecular potential, as approximated by Lennard-Jones. He also gives an integral expression for the second virial coefficients arising from statistical mechanics.

$$B = 2\pi N_A \int_0^\infty \left(1 - e^{-\frac{\Gamma}{kT}} \right) r^2 dr \qquad (8.1)$$

where N_A is Avogadro's number and $\Gamma(r)$ is the potential (Equation 7.1 or a variant). The third virial coefficient is much more complicated and involves a triple integral. In Section 5.3 on page 98, Prausnitz makes the following statement regarding a mixture of three components:

> The three second virial coefficients are functions only of the temperature; they are independent of density (or pressure) and what is most important, they are *independent of composition*. Since the second virial coefficient is concerned with interactions between *two* molecules, it can be rigorously shown that the second virial coefficient of a mixture is a *quadratic* function of the mole fractions, x_i and x_j.[16]

The quadratic expression for a two-component mixture is:

$$B_{mixture} = x_i^2 B_{ii} + 2x_i x_j B_{ij} + x_j^2 B_{jj} \qquad (8.2)$$

---

[16] The emphasis, "independent of composition," has been added and also the original variables y have been replaced with x, as it is customary, according to the IUPAC (International Union of Pure and Applied Chemistry), to use x when referring to mole fractions and y when referring to mass fractions; and so you will see on the Web, for instance, Wikipedia.

The general expression for an m-component mixture is:

$$B_{mixture} = \sum_{i=1}^{m}\sum_{j=1}^{m} x_i x_j B_{ij} \qquad (8.3)$$

The general expression for the third virial coefficient of the mixture is:

$$C_{mixture} = \sum_{i=1}^{m}\sum_{j=1}^{m}\sum_{k=1}^{m} x_i x_j x_k C_{ijk} \qquad (8.4)$$

The fugacity coefficient for component i, φ_i, follows from the virial expansion:

$$\ln \varphi_i = 2\rho \sum_{j=1}^{m} x_j B_{ij} + \frac{3\rho^2}{2} \sum_{j=1}^{m}\sum_{k=1}^{m} C_{ijk} - \ln Z_{mix} + ... \qquad (8.11)$$

Prausnitz states on page 101 (ibid) that the only limitation of this expression (Equation 6.11) is moderate densities and applies for any potential (e.g., ideal gas, hard sphere, or Lennard-Jones). For much of the remainder of his Chapter 5, Prausnitz discusses these various parameters for single-component fluids and various potentials.

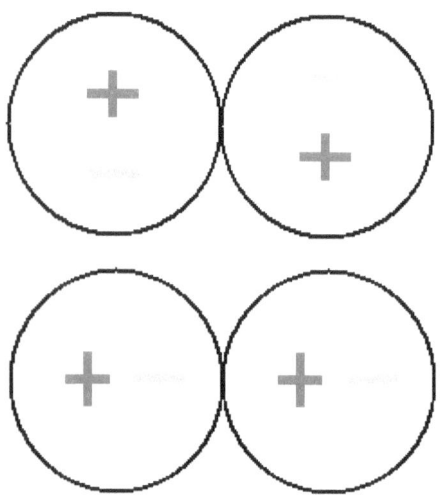

polar particle packing

In Section 5.12 on page 151, we see the first example of this relationship applied to a binary mixture that doesn't exhibit ideal behavior, that of carbon dioxide and n-butane (0.15/0.85 mole fractions), even up to considerable pressures (6000 psia or 400 atm). The ideal mixture results are shown along with the results obtained based on Equation 7.11. The calculations are in fairly good

agreement with data from Olds, Reamer, Sage, and Lacey.[17]   This is quite
promising, as this behavior approaches that of azeotropic.

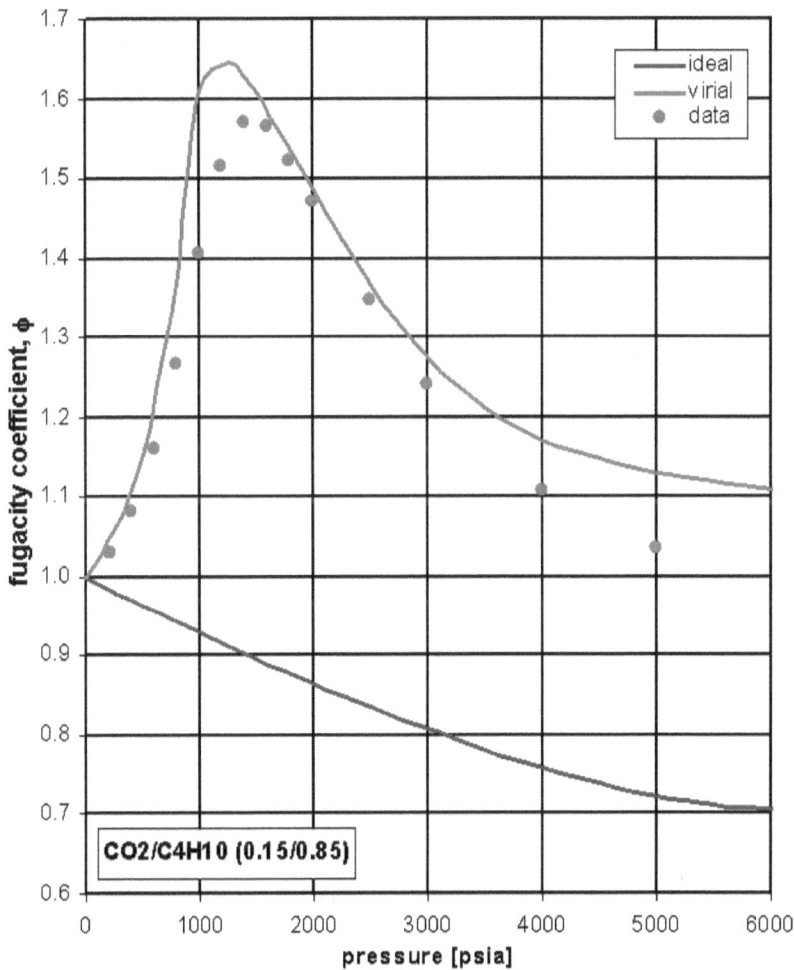

We next consider a mixture of nitrogen and ethylene. The data are from
Gunn, Chueh, and Prausnitz.[18] The pressures are quite large, up to 650 atm. The
agreement for this data set is also quite good and encouraging, especially

[17] Olds, R. H., Reamer, H. H., Sage, B. H., and Lacey, W. N., "The n-Butane–Carbon
   Dioxide System," Industrial & Engineering Chemistry, Vol. 41, No. 3, pp. 475-482,
   1949.
[18] Gunn, R. D., Chueh, P. L., and Prausnitz, J. M., "Prediction of Thermodynamic
   Properties of Dense Gas Mixtures Containing One or More of the Quantum Gases,"
   AIChE Journal, Vol. 12, pp. 937-941, 1966.

considering the very high pressures included in the comparison. The fact that the relationships work at three different mole fractions is also important. While these data are measured compressibilities and not fugacities, one does derive from the other.

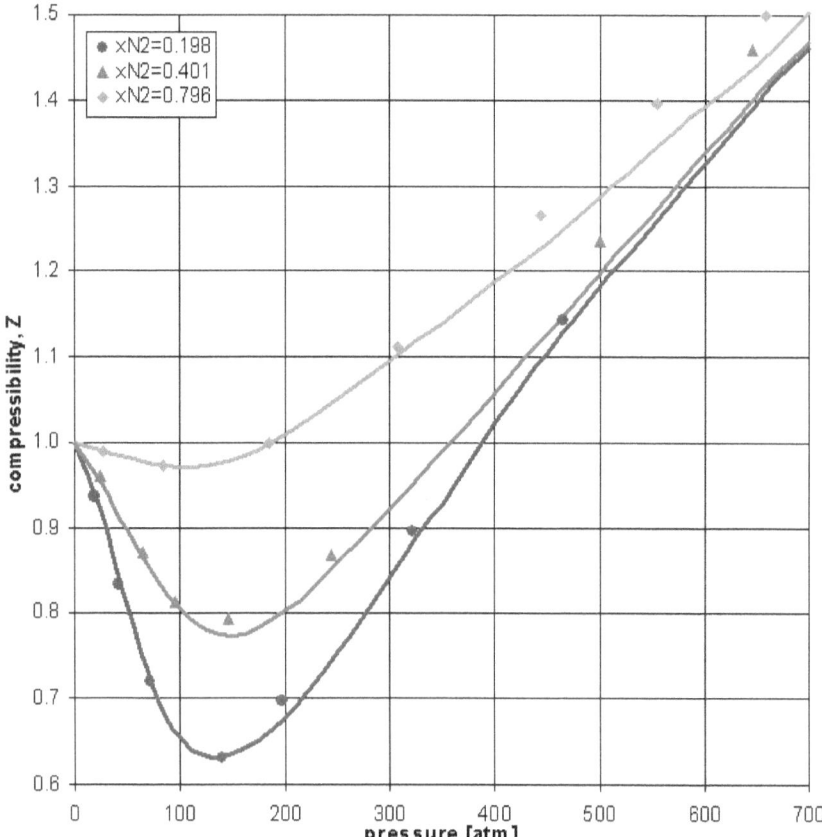

In this next data set, we consider sublimation, that is, vapor in equilibrium with solid at cryogenic temperatures. The data are for methane in hydrogen and come from the work of Hiza and Herring.[19] This data cover three orders of magnitude of the fugacity coefficient and pressures up to 160 atm, making this particularly interesting. The ideal mixture calculation is also shown, which is a single curve (independent of temperature), unlike calculations based on the virial coefficients, as described previously.

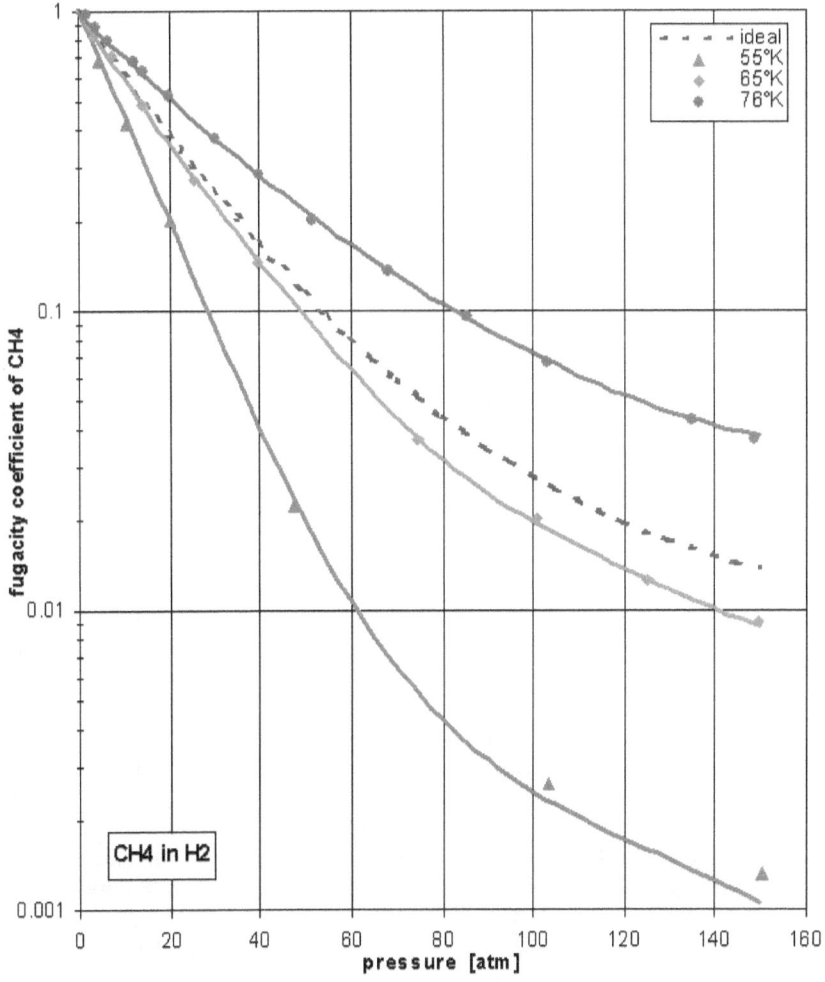

[19] Hiza, M. J. and Herring, R. N., "Solid-Vapor Equilibrium in the System Hydrogen-Methane," *International Advances in Cryogenic Engineering*, Plenum Press, New York, 1965.

In this next data set, we consider two gases (carbon dioxide in air) at cryogenic temperature (143°K) so that this measurement consisted of the mole fraction of CO_2 in air over dry ice (solid carbon dioxide). Two data sets are compared to the ideal calculation and virial: 1) Webster[20] and 2) Gratch.[21] Agreement with this data is important and also shows how far off the ideal calculations can be (orders of magnitude).

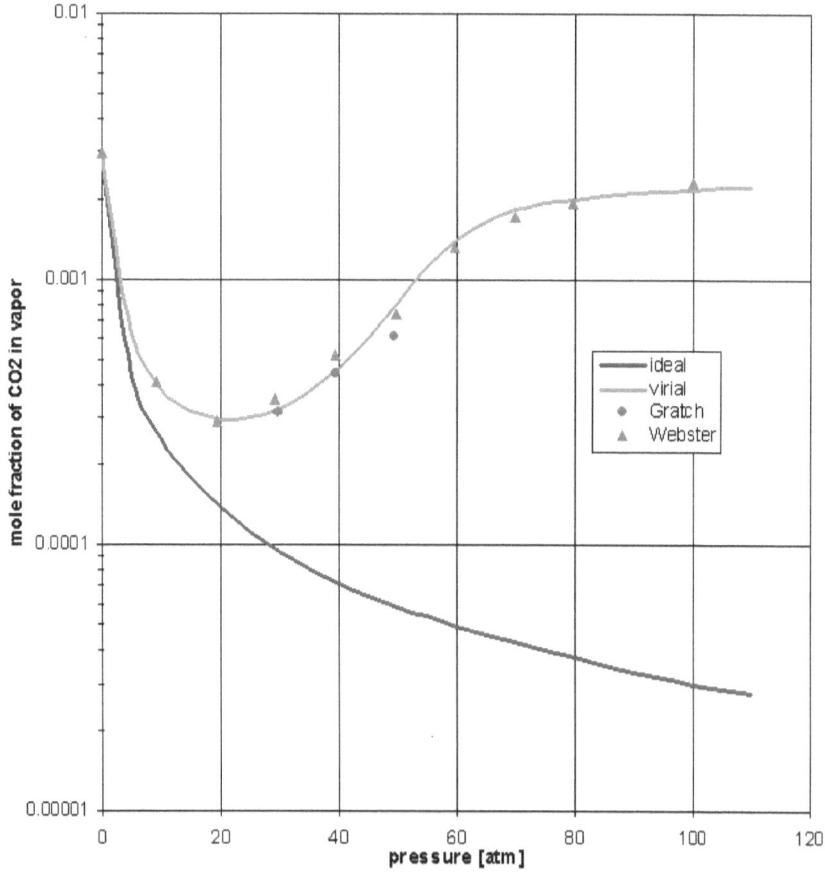

This next data set is for the equilibrium mole fraction of naphthalene in the vapor state above naphthalene in the liquid state under pressure of ethylene gas over the range of 12°C to 35°C (285°K to 308°K). The free energies must be in balance in order for the concentration to stabilize. The data are from Diepen &

[20] Webster, T. J., "The Influence of Pressure on the Equilibrium between Carbon Dioxide and Air," *Proceedings of the Royal Society* (London), Vol. A214, No. 61, 1952.
[21] Gratch, S., "Project G-9A," University of Pennsylvania Thermodynamics Research Laboratory, 1945-1946.

Scheffer.[22] The difference between actual and ideal is so great that the vertical scale is separated by four orders of magnitude.

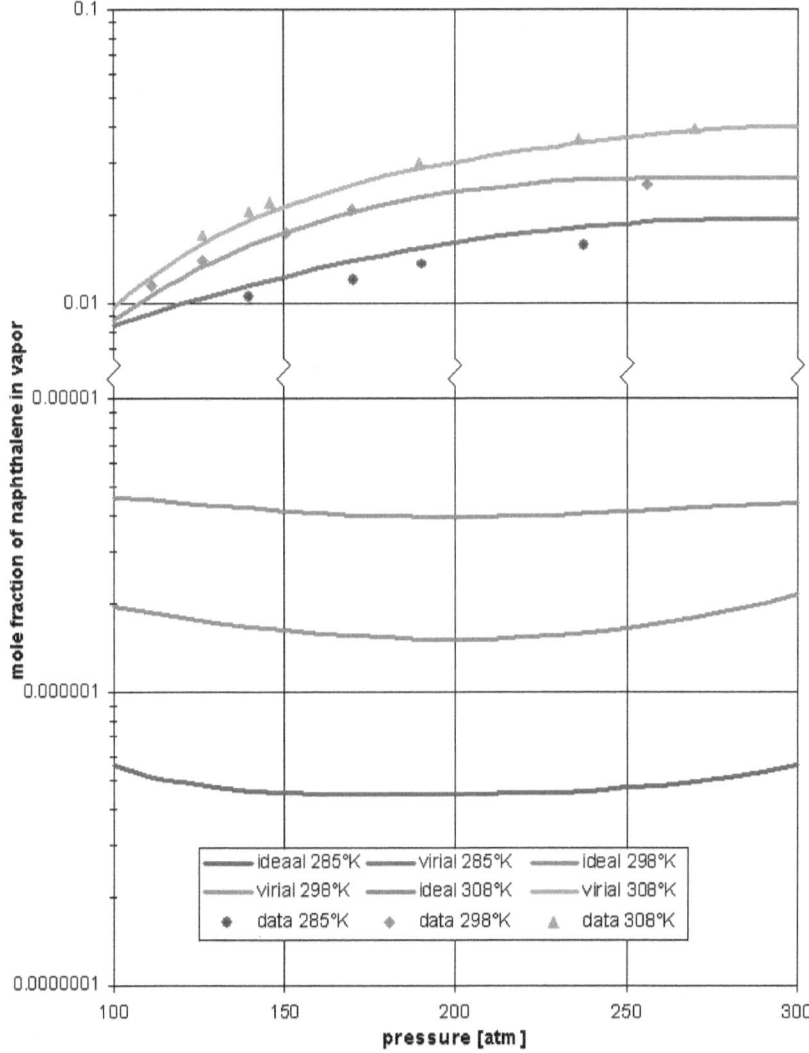

The final data set we will consider in this chapter is reminiscent of the water and 2-propanol system in Chapter 1. Here we consider ethanol over isooctane,

[22] Diepen, G. A. and Scheffer, F. E., "The Solubility of Naphthalene in Supercritical Ethylene," *Journal of the American Chemical Society*, Vol. 70, No. 12, pp. 4085-4089, 1948.

which is even more pronounced, yet the virial approach fits the data quite well, indicating that we have at last found a suitable approach to handle these unusual fluid mixtures. The data are from Kretschmer.[23] This particular data and calculation are of some interest in the area of biofuels, as it relates the mixing and vaporization of ethanol and isooctane [note: n-octane has an anti-knock rating of 0, while isooctane has a rating of 100, making it the relevant standard for motor fuels].

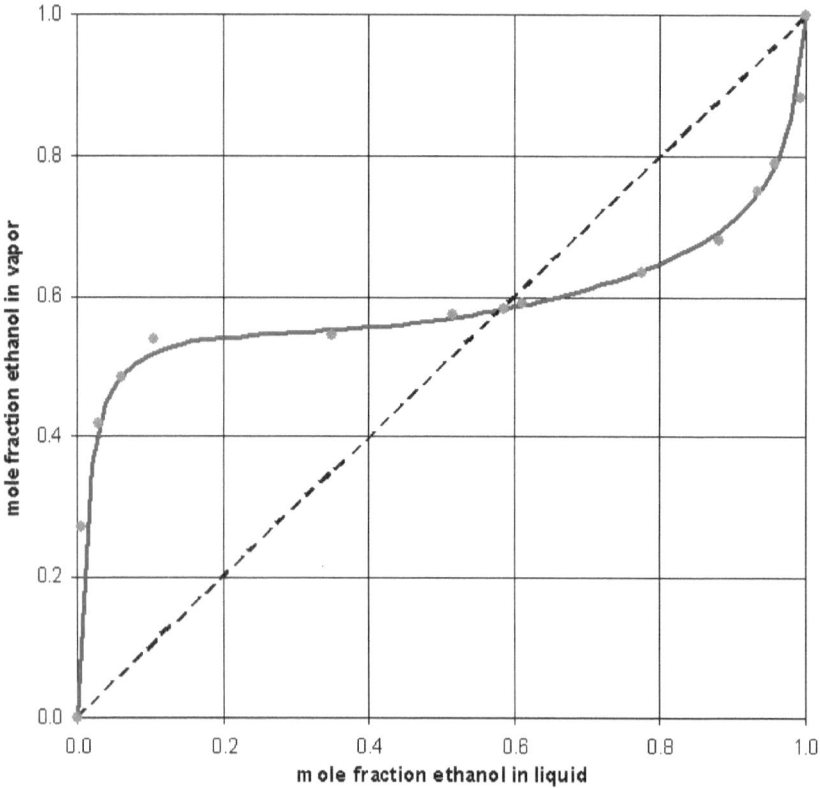

---

[23] Kretschmer, C. B., "Densities and Liquid-Vapor Equilibria of the System: Ethanol-Isooctane (2,2,4-trimethylpentane) between O and 50 Degrees," Journal of the American Chemical Society, Vol. 70, 1948.

Chapter 9. Transport Properties and Sizing

Temperature ranges and flow rates are not the only factors in selecting a working fluid. Transport properties also impact the size and cost of a system. We already discussed the issue of specific volume, that is, if the specific volume is very large (very small density), then the compressor in a refrigeration cycle will be relatively large. While we might welcome a large turbine to expand a rarefied vapor, the condenser will also be relatively large. Power plant steam condensers can be the size of a house.

Pressure Drop

We first consider pressure drop. Most often this is for turbulent flow in a pipe. The basic relationship is called the Darcy-Weisbach[24,25] equation:

$$\Delta P = f\left(\frac{L}{d}\right)\left(\frac{\rho V^2}{2g}\right) \tag{9.1}$$

The coefficient, f, is the friction factor. The most common representation of the friction factor is called the Moody[26] Chart:

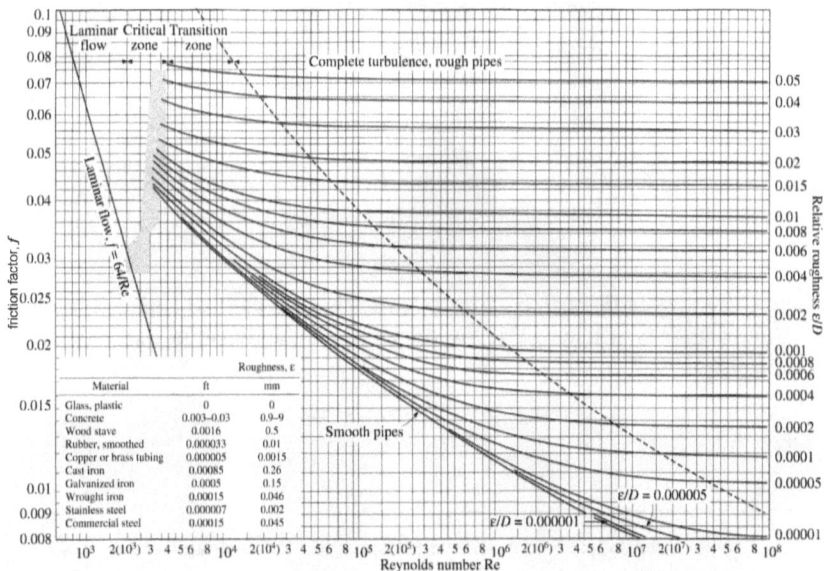

---

[24] Henry Philibert Gaspard Darcy (1803–1858) French engineer who made important advances in the study of hydraulics and flow in porous media.
[25] Julius Ludwig Weisbach (1806-1871) German mathematician and engineer.
[26] Lewis Ferry Moody (1880–1953) American engineer and inventor; the first Professor of Hydraulics in the School of Engineering at Princeton.

There is also a Fanning[27] friction factor, which is equal to one-fourth of the Darcy-Weisbach friction factor. Charts are not always clear on this and confusion is common. The easiest way to tell the difference between the two is to look at the laminar line on the left side of the chart. If it says $f=Re/64$, then it's Darcy-Weisbach. If it says $f=Re/16$, then it's the Fanning.

Convective Heat Transfer

The difference between laminar and turbulent convection is significant. For this reason, designers size equipment, such as pipes, to assure turbulent flow. The basic relationship for turbulent convective heat transfer is called Reynolds[28] Analogy.

$$St = \frac{f}{2} \tag{9.2}$$

where St is the Stanton[29] number, which can be related to the Nusselt[30], Reynolds, and Prandtl[31] numbers:

$$St = \frac{Nu}{RePr} = \frac{h}{\rho CV} \tag{9.3}$$

$$Re = \frac{\rho Vd}{\mu} \tag{9.4}$$

$$Pr = \frac{\mu C}{k} \tag{9.5}$$

$$Nu = \frac{hd}{k} \tag{9.6}$$

where C is the specific heat, d is the diameter, k is the thermal conductivity, h is the heat transfer coefficient, V is the velocity, μ is the dynamic viscosity, and ρ is the density.

---

[27] John Thomas Fanning (1837–1911) American architect and hydraulic engineer who studied pressure loss in pipes.

[28] Osborne Reynolds (1842–1912) pioneer in the field of fluid dynamics, conducting studies of boilers and condensers.

[29] Thomas Edward Stanton (1865-1888) British engineer; studied under Osborne Reynolds, jointly conducted many experiments.

[30] Ernst Kraft Wilhelm Nußelt (1882–1957) German engineer and professor, who performed many studies advancing the science of heat transfer.

[31] Ludwig Prandtl (1875–1953) German engineer and aerodynamicist; pioneer in the field of aeronautics.

Equation 9.2 was refined to arrive at the correlation of Dittus-Boelter:[32]

$$Nu = 0.023\,Re^{0.8}\,Pr^{n} \tag{9.7}$$

where n=0.4 for heating and n=0.3 for cooling. The range of validity for Equation 9.7 is $Re \geq 10^5$ and $0.6 \leq Pr \leq 160$.

Condensation Heat Transfer

While there might be a few batch processing or experimental applications for condensation on a vertical plate, such designs are not effective for industrial processes. Most often condensation is effected on the outside of horizontal tubes within a shell and so we will consider this mode of heat transfer as we evaluate the various fluids. The most common correlation for film condensation on a tube bank was developed by Nusselt:

$$h = 0.725\left[\frac{\rho_L(\rho_L - \rho_V)g\,h_{FG}\,k_L^3}{\mu_L d\Delta T}\right]^{\frac{1}{4}} \tag{9.8}$$

where subscripts L and V denote liquid and vapor, h_{FG} is the latent heat of vaporization, and ΔT is the temperature difference across the condensing film.

Boiling Heat Transfer inside Tubes

There are several regimes for boiling heat transfer from occasional bubbles to rapid vaporization and drying-out of the surface, all proportional to temperature difference. After the surface dries out (consistently or sporadically), the heat transfer diminishes significantly, as the thermal conductivity of the vapor (now in contact with the surface) is much lower than that of the liquid. The peak (near but before dry-out) is called *critical* (not to be confused with critical temperature and pressure, which are thermodynamic properties). We will not consider temperature differences anywhere near this critical, as such are not the most efficient conditions to provide refrigeration or produce power.

While there are many cases of *pool* boiling (think: a pot of water on the stove) these are rarely used for industrial processes, as they are not efficient uses of expensive materials. What we will consider is *nucleate* boiling (think: bubbles) inside vertical tubes. This is what happens in a large coal-fired boiler, the evaporator in a heat recovery steam generator as part of a combined cycle power plant, and in the air conditioning evaporator located behind the dashboard of a car.

---

[32] Dittus, F. W. and Boelter, L. M. K., "Heat Transfer in Automobile Radiators of the Tubular Type," *Publications in Engineering*, University of California, Berkeley, Vol. 2, 1930.

For our consideration of fluid comparison, we will use the correlation of Davis & David:[33]

$$Nu = 0.06\left(\frac{\rho_L}{\rho_V}\right)^{0.28} Re^{0.87} Pr^{0.4} \tag{9.9}$$

As we have seen for turbulent convection, viscosity and thermal conductivity significantly impact heat transfer coefficients. Density and surface tension typically have a lesser impact. Surface tension often comes into play with phase change (condensation or boiling), although not in the correlation selected here for our comparison of fluids. Low heat transfer coefficients (convective, evaporating, or condensing) translate directly into large heat exchange areas and greater cost. See surface_tension.xls for this graph:

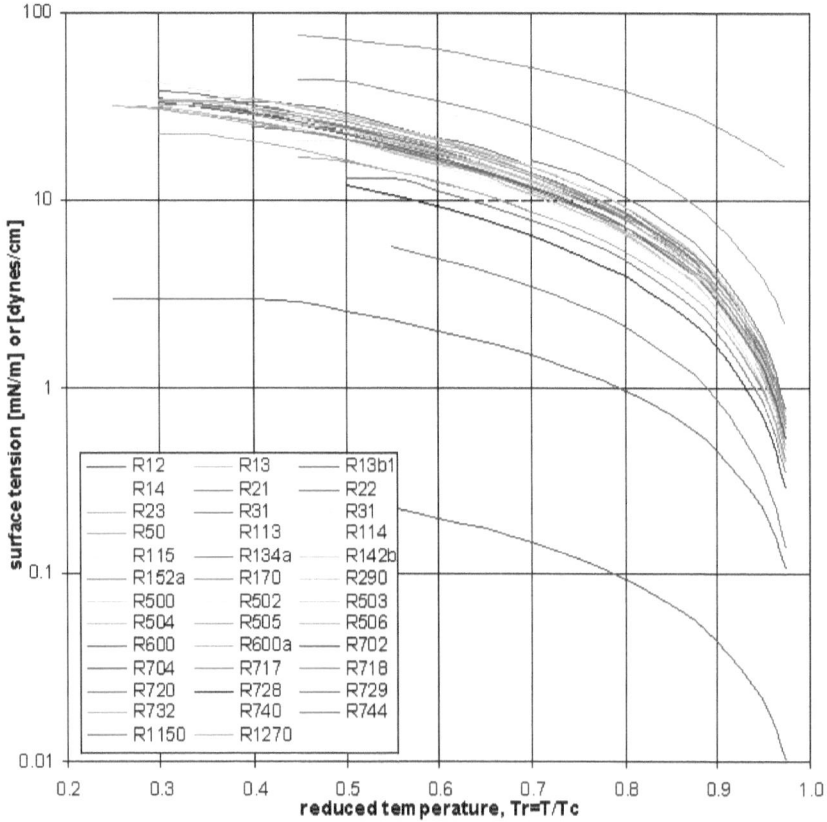

---

[33] Davis, E. J. and David, M. M., "Two-Phase Gas-Liquid Convection Heat Transfer," *Industrial & Engineering Chemistry Fundamentals*, Vol. 3, pp. 111-118, 1964.

The dynamic and kinematic viscosity is also available in the Excel AddIn. You will find this plot in spreadsheet viscosity.xls on the dynamic tab. The X-axis is reduced temperature, $T_R = T/T_C$. The curves are reduced pressure, $P_R = P/P_C$, isobars (lines of constant pressure).

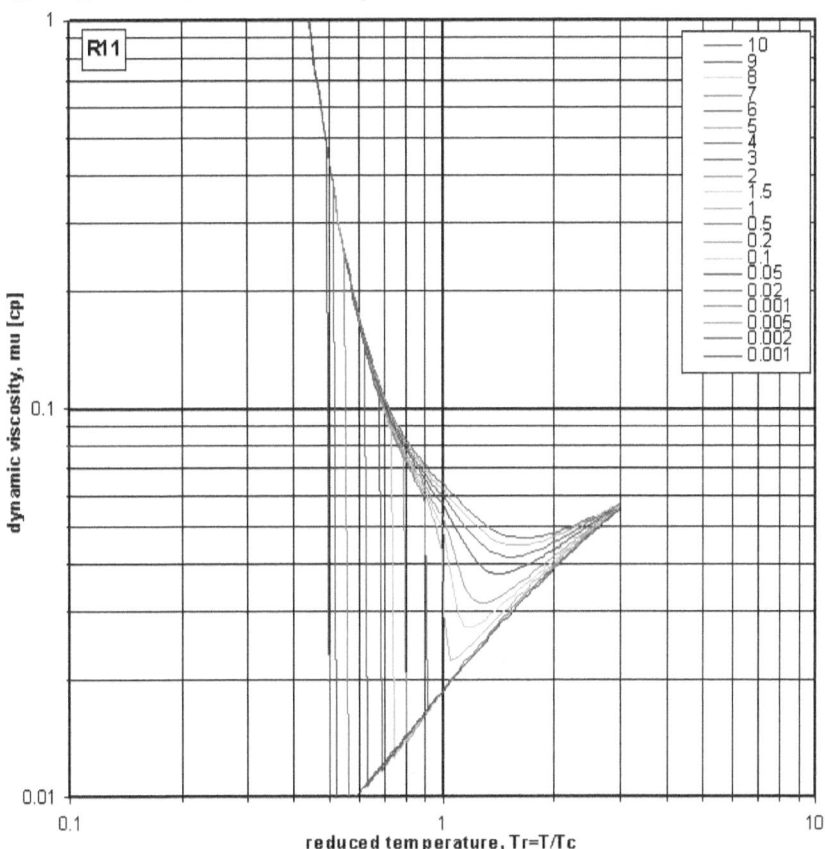

This plot of kinematic viscosity can be found in the same spreadsheet on the other tab. The curves are the same isobars. You can select the fluid and units.

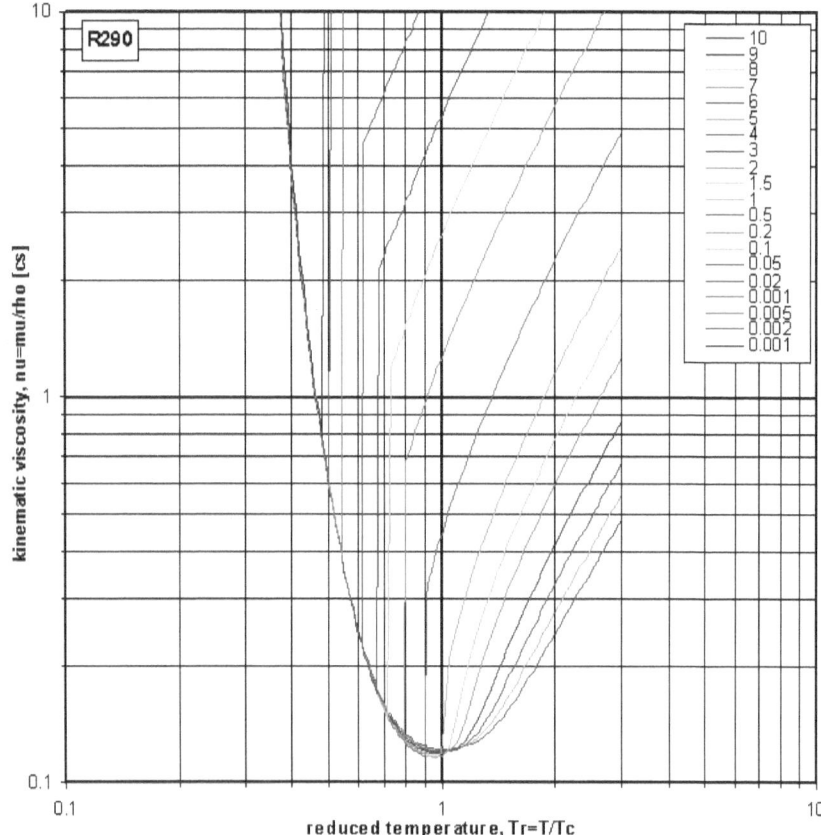

The constant pressure (C_P) and constant volume (C_V) specific heats are included in the Excel AddIn. You will find this next figure in spreadsheet specific_heat.xls on the C_P tab. Note that the specific heat becomes unbounded (blows up) at the saturation point because the enthalpy changes by h_{FG} at constant pressure from liquid to vapor with no change in temperature ($\Delta T=0$).

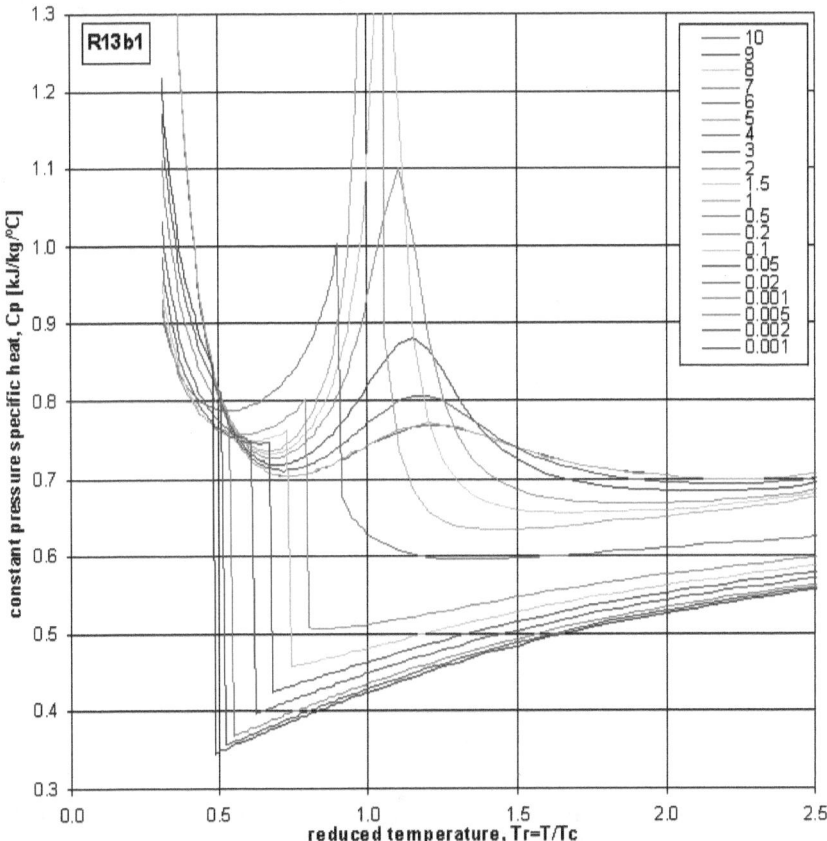

The constant volume specific heat (C_V) is in the same spreadsheet on the other tab.

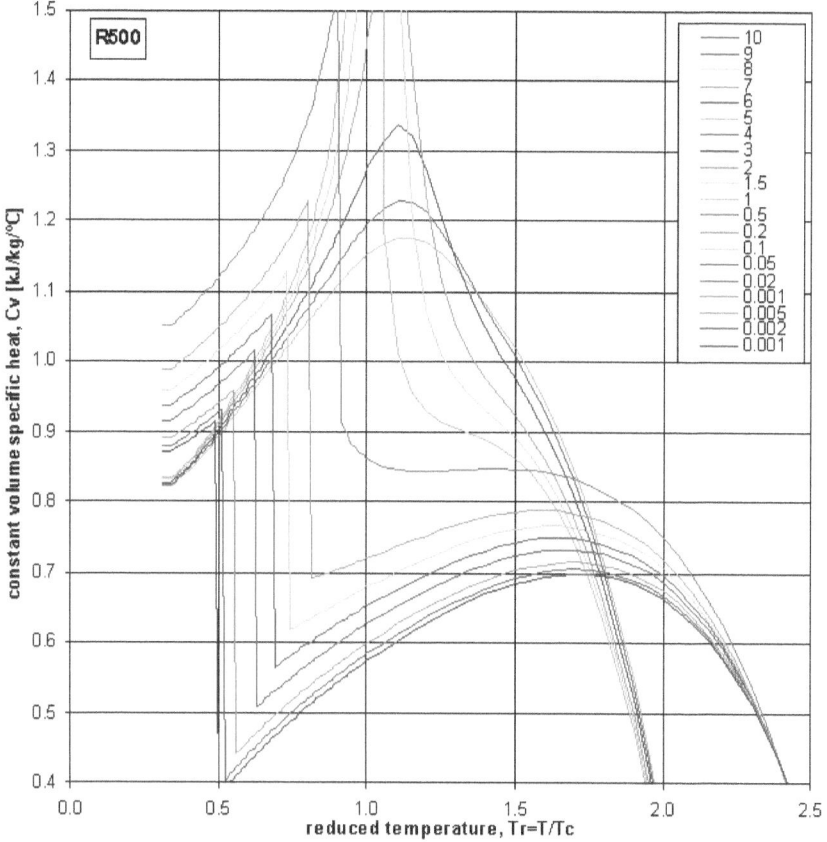

With all of the necessary thermodynamic and transport properties available in the Excel AddIn and the typical relationship (Equation 9.7), we can now compare the fluids on this basis. We will consider two sets of temperatures, both from Table 4.1: the lower temperature for cooling and the higher one for heating. The results for cooling are shown first in the table on the next page. The calculations can be found in spreadsheet heat_transfer.xls. The typical tube velocity (2.5 m/s) and tube diameter (1.0 cm) can be selected. The units can also be selected, but some care is needed to make sure the calculations work out and there are many combinations of units available. The refrigerants have been sorted in order of decreasing heat transfer coefficient (think: more surface area and material required). Several cells are highlighted in red, indicating that the Reynolds number is too small ($<10^4$).

Table 9.1 Convective Cooling Comparison

convective heat transfer for cooling 2.5 m/s velocity in a 1.0 cm diameter tube											
refrigerant	T	P	r	Cp	m	k	Pr	Re	Nu	h	relative
number	°C	kPa	kg/m³	kJ/kg/°C	cp	W/m/°C	-	-	-	W/m²/°C	-
R31	-152.8	1.8E-04	1826	9.999	0.010	0.071	1	4.6E+06	5423	9999	329%
R720	-243.6	4.0E+02	1164	2.064	0.021	0.127	0	1.4E+06	1370	9999	320%
R729	-193.3	1.6E+02	834	2.742	0.042	0.130	1	4.9E+05	797	9999	261%
R744	-51.6	1.3E+03	1162	2.001	0.084	0.129	1	3.5E+05	672	8645	218%
R728	-205.0	5.7E+01	846	2.036	0.062	0.143	1	3.4E+05	590	8458	213%
R740	-184.3	2.3E+02	1386	1.149	0.057	0.097	1	6.1E+05	868	8456	213%
R717	-72.7	1.8E+01	728	4.131	0.352	0.426	3	5.2E+04	196	8345	211%
R702	-254.2	1.3E+02	72	8.067	0.010	0.265	0	1.8E+05	259	6858	173%
R718	5.0	1.7E+00	1000	4.181	1.520	0.572	11	1.6E+04	112	6393	161%
R14	-146.1	4.6E+01	1696	0.857	0.111	0.058	2	3.8E+05	781	4519	114%
R23	-150.0	2.8E-01	2100	9.999	3.341	0.129	258	1.6E+04	277	3583	90%
R114	12.9	2.8E+02	1492	0.997	0.098	0.041	2	3.8E+05	869	3580	90%
R50	-177.5	4.2E+01	445	3.420	0.208	0.108	7	5.3E+04	245	2658	67%
R134a	-98.3	1.3E+00	1578	1.187	0.536	0.075	8	7.4E+04	342	2566	65%
R113	-16.4	1.3E+01	1660	0.890	0.285	0.050	5	1.5E+05	505	2542	64%
R504	-138.6	3.0E-01	1833	1.428	1.470	0.093	23	3.1E+04	231	2145	54%
R503	-146.1	1.2E+00	1621	1.090	0.922	0.079	13	4.4E+04	255	2025	51%
R115	-94.4	6.8E+00	1729	0.856	0.338	0.039	7	1.3E+05	512	1982	50%
R502	-101.1	5.8E+00	1546	0.969	0.508	0.051	10	7.6E+04	364	1871	47%
R732	-213.8	1.2E+00	1285	1.688	2.291	0.134	29	1.4E+04	131	1759	44%
R1150	-146.1	6.4E+00	625	2.400	0.630	0.100	15	2.5E+04	170	1707	43%
R40	-92.7	3.1E+00	1106	1.195	0.978	0.092	13	2.8E+04	180	1644	41%
R21	-68.2	2.6E+00	1571	0.898	0.815	0.060	12	4.8E+04	272	1630	41%
R152a	-113.6	2.5E-01	1184	1.541	1.415	0.090	24	2.1E+04	171	1539	39%
R704	-270.7	1.7E+01	145	2.274	1.233	0.467	6	2.9E+03	23	1097	28%
R506	-109.4	1.3E-01	1591	1.335	4.690	0.078	80	8.5E+03	119	932	24%
R505	-118.2	5.6E-01	1529	0.610	4.491	0.073	38	8.5E+03	95	691	17%
R142b	-125.4	1.6E-02	1440	1.123	8.251	0.066	141	4.4E+03	83	544	14%
R11	-105.5	2.6E-02	1759	0.784	14.134	0.044	249	3.1E+03	75	333	8%
R1270	-146.1	3.3E-02	723	2.092	18.128	0.095	399	1.0E+03	35	330	8%
R500	-140.0	2.2E-02	1562	0.844	14.711	0.049	254	2.7E+03	66	324	8%
R22	-152.4	2.3E-03	1708	1.068	46.675	0.059	852	9.1E+02	41	238	6%
R600	-133.3	3.3E-03	730	2.010	54.367	0.102	1071	3.4E+02	20	200	5%
R13b1	-162.8	1.2E-03	2299	1.784	99.999	0.030	6014	5.7E+02	50	150	4%
R170	-177.8	7.9E-03	646	2.365	99.999	0.091	2606	1.6E+02	14	129	3%
R600a	-154.4	1.6E-04	736	1.745	99.999	0.087	2009	1.8E+02	15	127	3%
R12	-152.1	1.5E-03	1816	0.864	99.999	0.040	2175	4.5E+02	31	122	3%
R290	-182.6	2.3E-06	728	1.938	99.999	0.079	2440	1.8E+02	15	122	3%
R13	-176.1	2.6E-03	1848	0.800	99.999	0.039	2069	4.6E+02	31	119	3%

The calculations for convective heating are shown in this next table. All of the Reynolds numbers are in range and there are considerable differences.

Table 9.2 Convective Heating Comparison

refrigerant number	T °C	P kPa	r kg/m³	Cp kJ/kg/°C	m cp	k W/m/°C	Pr -	Re -	Nu -	h W/m²/°C	relative -
						convective heat transfer for heating 2.5 m/s velocity in a 1.0 cm diameter tube					
R718	369.1	41650	630	6.485	0.074	0.480	1.0	2.1E+05	421	20232	159%
R720	-233.7	2684	951	2.774	0.012	0.112	0.3	2.0E+06	1755	19647	159%
R717	90.0	10159	498	8.203	0.050	0.336	1.2	2.5E+05	508	17047	159%
R732	-127.0	7185	815	2.329	0.017	0.132	0.3	1.2E+06	1157	15298	159%
R728	-152.0	5165	574	2.774	0.013	0.117	0.3	1.1E+06	1122	13150	159%
R729	-145.5	5700	585	2.813	0.013	0.106	0.3	1.1E+06	1154	12225	159%
R740	-127.5	7777	974	1.626	0.018	0.079	0.4	1.3E+06	1368	10751	159%
R504	34.5	4521	1288	1.638	0.054	0.090	1.0	6.0E+05	955	8619	137%
R503	-3.5	4874	1169	1.452	0.039	0.081	0.7	7.4E+05	1032	8317	132%
R506	69.9	2473	1320	1.496	0.070	0.098	1.1	4.7E+05	812	7976	126%
R744	26.0	13166	852	2.479	0.050	0.097	1.3	4.3E+05	789	7676	105%
R505	45.1	2138	1253	1.186	0.063	0.086	0.9	5.0E+05	794	6823	94%
R14	-53.6	5906	1136	1.329	0.026	0.046	0.8	1.1E+06	1425	6499	89%
R23	13.6	7163	1014	1.862	0.049	0.068	1.3	5.2E+05	935	6392	88%
R134a	37.5	1896	1165	1.562	0.058	0.069	1.3	5.0E+05	909	6257	86%
R600	37.0	694	559	2.600	0.063	0.131	1.3	2.2E+05	467	6093	84%
R152a	54.3	2614	826	2.144	0.058	0.084	1.5	3.5E+05	713	5999	82%
R1270	33.1	2807	497	3.025	0.047	0.107	1.3	2.7E+05	547	5849	80%
R40	123.9	9840	720	2.269	0.049	0.080	1.4	3.7E+05	717	5748	79%
R22	42.4	3250	1132	1.482	0.059	0.065	1.3	4.8E+05	876	5730	79%
R600a	17.2	554	561	2.464	0.064	0.123	1.3	2.2E+05	464	5721	78%
R170	18.1	7241	385	3.531	0.035	0.101	1.2	2.8E+05	550	5535	76%
R290	27.6	2026	492	2.946	0.061	0.121	1.5	2.0E+05	452	5468	75%
R500	25.5	1560	1265	1.299	0.074	0.066	1.5	4.3E+05	825	5402	74%
R702	-245.0	1181	61	9.999	0.010	0.204	0.5	1.5E+05	261	5317	73%
R50	-87.6	7693	294	4.862	0.026	0.091	1.4	2.8E+05	584	5305	73%
R142b	35.2	913	1087	1.381	0.075	0.075	1.4	3.6E+05	709	5296	73%
R502	20.2	1953	1297	1.264	0.066	0.056	1.5	4.9E+05	925	5151	71%
R21	109.3	3202	1139	1.313	0.063	0.057	1.5	4.5E+05	861	4901	67%
R1150	4.2	9024	388	3.360	0.035	0.084	1.4	2.8E+05	578	4875	67%
R12	39.0	1868	1266	1.083	0.064	0.049	1.4	4.9E+05	911	4504	62%
R13b1	18.0	2758	1708	0.992	0.074	0.040	1.8	5.8E+05	1125	4483	61%
R31	33.4	852	1315	1.344	0.142	0.073	2.6	2.3E+05	601	4375	60%
R13	7.4	4737	1122	1.211	0.048	0.042	1.4	5.8E+05	1036	4368	60%
R113	109.7	1088	1344	1.050	0.070	0.047	1.6	4.8E+05	921	4323	59%
R11	75.0	924	1351	0.962	0.086	0.053	1.6	3.9E+05	787	4176	57%
R115	10.0	1191	1361	1.096	0.058	0.037	1.7	5.8E+05	1108	4122	57%
R114	85.6	2081	1251	1.220	0.060	0.036	2.0	5.2E+05	1065	3842	53%
R704	-268.2	354	123	7.748	0.085	0.032	20.7	3.6E+04	253	806	11%

The low temperature properties of R31 are quite remarkable, but the high temperature properties are mundane. This is an important consideration when selecting working fluids: it must perform well in each process comprising the entire cycle. Excluding R31, the range of heat transfer coefficients for cooling is only twice the range for heating. Still, there is such a great difference that you can see why some fluids are rarely used. While there is no clear pattern for the azeotropes in either table, they are roughly in the middle of Table 9.1 and slightly above the middle in Table 9.2.

Table 9.3 shows the results from the two preceding tables plus a composite score for both convective cooling and heating (harmonic). In this table we have some clear winners (the 700 series). The convective heat transfer advantage of the 700 series is quite pronounced when viewed graphically, as on the page following Table 9.3. The azeotropes are distributed with none at the top or bottom of the list. We notice that the most common refrigerants (R11-R22) are at the bottom of the list, which means that as far as this comparison is concerned, there is opportunity for improvement.

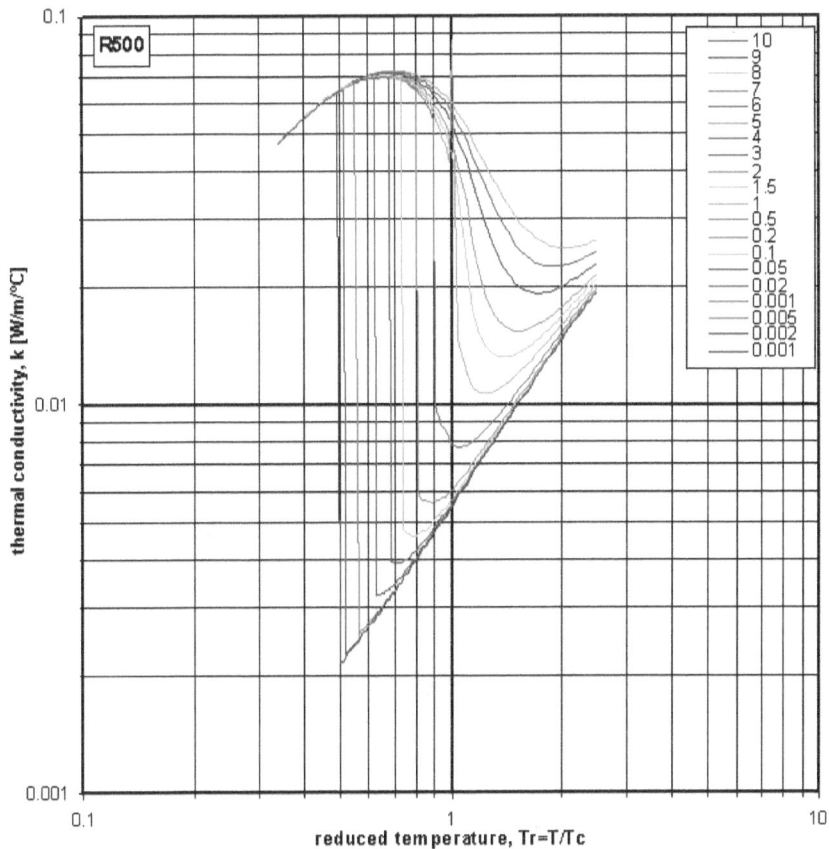

Table 9.3 Composite Convective Comparison

refrigerant	cooling	heating	composite
R720	440%	270%	167%
R717	211%	234%	111%
R729	261%	168%	102%
R718	161%	278%	102%
R728	213%	180%	98%
R740	213%	147%	87%
R744	218%	105%	71%
R31	969%	60%	57%
R702	173%	73%	51%
R14	114%	89%	50%
R23	90%	88%	45%
R504	54%	118%	37%
R134a	65%	86%	37%
R732	44%	210%	37%
R503	51%	114%	35%
R50	67%	73%	35%
R114	90%	53%	33%
R113	64%	59%	31%
R502	47%	71%	28%
R40	41%	79%	27%
R115	50%	57%	27%
R152a	39%	82%	26%
R1150	43%	67%	26%
R21	41%	67%	26%
R506	24%	109%	19%
R505	17%	94%	15%
R142b	14%	73%	12%
R704	28%	11%	8%
R1270	8%	80%	8%
R500	8%	74%	7%
R11	8%	57%	7%
R22	6%	79%	6%
R600	5%	84%	5%
R13b1	4%	61%	4%
R170	3%	76%	3%
R600a	3%	78%	3%
R290	3%	75%	3%
R12	3%	62%	3%
R13	3%	60%	3%

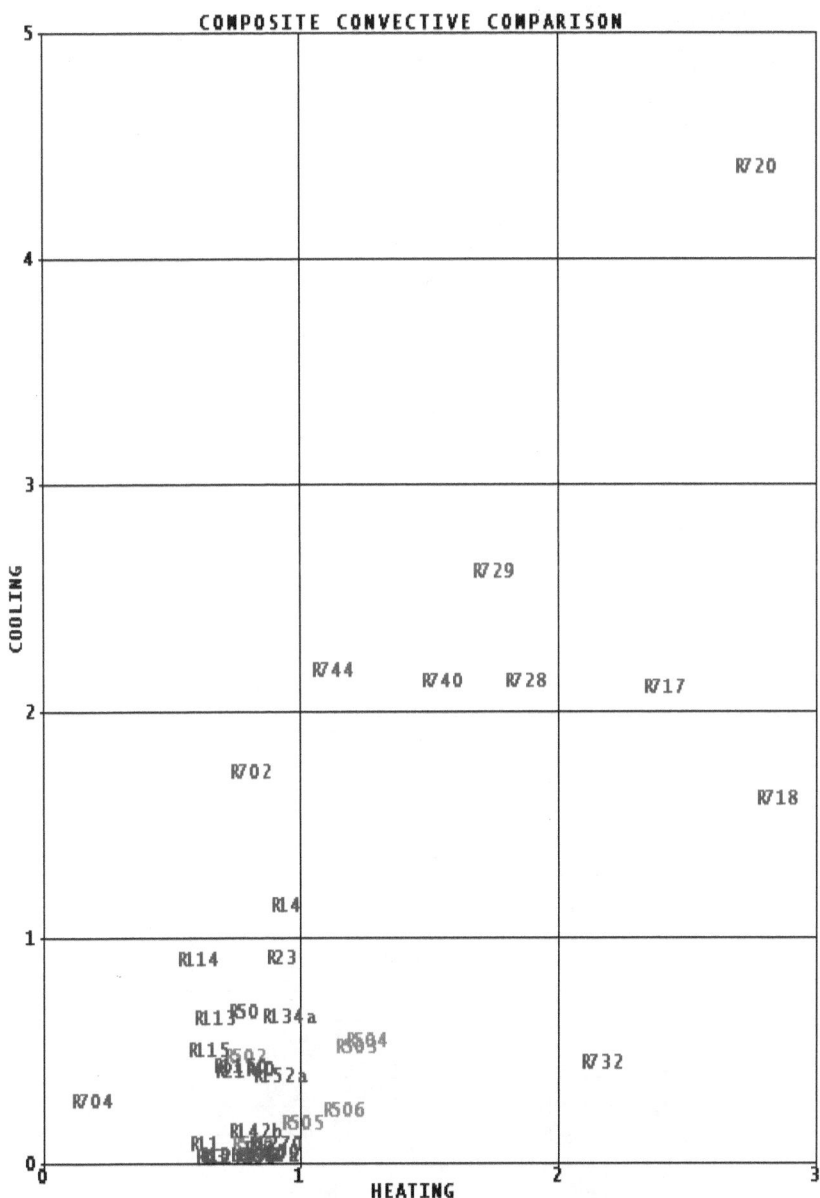

Condensation Heat Transfer Coefficient Comparison

We perform this comparison using Equation 9.8, again in spreadsheet heat_transfer.xls on the condensation tab. The user-defined variables this time include tube diameter (2.5 cm) and temperature difference (5°C).

Table 9.3 Condensation Heat Transfer Coefficient Comparison

refrigerant number	T °C	r_L kg/m³	r_V kg/m³	h_{FG} kJ/kg	m_L cp	k_L W/m/°C	h W/m²/°C	relative -

refrigerant number	T °C	r_L kg/m³	r_V kg/m³	h_{FG} kJ/kg	m_L cp	k_L W/m/°C	h W/m²/°C	relative -
R718	5.0	1000	0.007	2478	1.520	0.572	9016	437%
R717	-72.7	728	0.092	1474	0.352	0.426	7799	378%
R31	-152.8	1826	0.000	898	0.010	0.071	6929	336%
R744	-51.6	1160	16.845	342	0.084	0.129	3974	193%
R720	-243.6	1163	17.900	83	0.021	0.127	3921	190%
R729	-193.3	834	3.495	228	0.042	0.130	3650	177%
R23	-150.0	2100	0.010	2812	3.341	0.129	3632	176%
R728	-205.0	846	1.433	210	0.062	0.143	3524	171%
R740	-184.3	1386	6.519	160	0.057	0.097	3226	156%
R702	-254.2	72	0.894	454	0.010	0.265	3117	151%
R14	-146.1	1696	1.945	143	0.111	0.058	1987	96%
R50	-177.5	445	0.429	538	0.208	0.108	1938	94%
R134a	-98.3	1578	0.047	261	0.536	0.075	1825	89%
R504	-138.6	1833	0.011	252	1.470	0.093	1781	86%
R40	-92.7	1106	0.052	463	0.978	0.092	1761	85%
R732	-213.8	1285	0.040	239	2.291	0.134	1735	84%
R1150	-146.1	625	0.085	539	0.630	0.100	1643	80%
R503	-146.1	1621	0.048	212	0.922	0.079	1598	78%
R152a	-113.6	1184	0.006	402	1.415	0.090	1582	77%
R114	12.9	1491	10.560	133	0.098	0.041	1460	71%
R113	-16.4	1660	0.562	164	0.285	0.050	1447	70%
R21	-68.2	1571	0.080	274	0.815	0.060	1404	68%
R502	-101.1	1546	0.228	194	0.508	0.051	1280	62%
R506	-109.4	1591	0.004	313	4.690	0.078	1152	56%
R115	-94.4	1729	0.356	144	0.338	0.039	1125	55%
R505	-118.2	1529	0.023	207	4.491	0.073	972	47%
R704	-270.7	145	1.866	23	1.233	0.467	964	47%
R142b	-125.4	1440	0.001	284	8.251	0.066	813	39%
R1270	-146.1	723	0.001	539	18.128	0.095	733	36%
R600	-133.3	730	0.000	492	54.367	0.102	577	28%
R500	-140.0	1562	0.001	250	14.711	0.049	569	28%
R11	-105.5	1759	0.001	226	14.134	0.044	554	27%
R22	-152.4	1708	0.000	300	46.675	0.059	534	26%
R170	-177.8	646	0.000	589	99.999	0.091	447	22%
R600a	-154.4	736	0.000	477	99.999	0.087	437	21%
R290	-182.6	728	0.000	558	99.999	0.079	423	21%
R12	-152.1	1816	0.000	215	99.999	0.040	313	15%
R13	-176.1	1848	0.000	192	99.999	0.039	301	15%
R13b1	-162.8	2299	0.000	186	99.999	0.030	273	13%

The order is quite similar to the two previous ones, with the 700 series fluids at the top and the azeotropes scattered throughout the mid range. The most common refrigerants are not all at the bottom, for instance R23 is in the upper quartile. R12 is still near the bottom.

Convective Boiling Heat Transfer Coefficient Comparison

We perform this comparison using Equation 9.9, again in spreadsheet heat_transfer.xls on the evaporation tab. The user-defined variables this time include tube velocity (2.5 m/s), tube diameter (2.5 cm), and temperature difference (5°C). The graphical comparison for phase change shown below is even more striking than for convection.

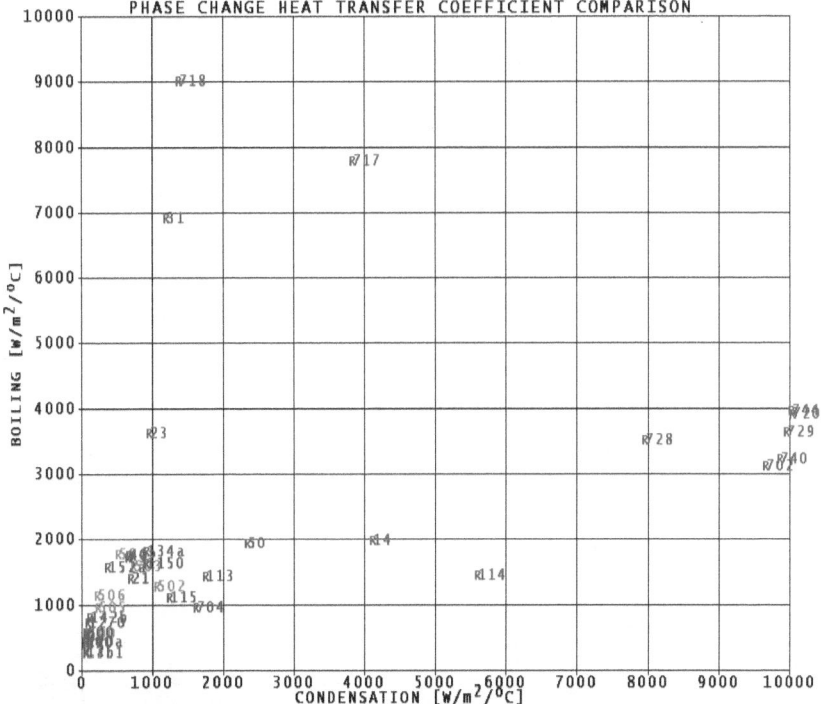

The 700 series is again the most outstanding. These all have relatively low molecular weight and are non-flammable. The last comparison we will consider is an overall comparison of all three heat transfer coefficients and molecular weight, which is on the page following Table 9.4.

Table 9.4 Convective Boiling Heat Transfer Coefficient Comparison

refrigerant number	T °C	rL kg/m³	rv kg/m³	hFG kJ/kg	CPL kJ/kg/°C	mL cp	kL W/m/°C	Pr -	Re -	Nu -	h W/m²/°C	relative -
boiling heat transfer inside 2.5 cm diameter tubes with 5.0 °C dT												
R720	-243.6	1163	17.900	83	2.064	0.021	0.127	0	3.5E+06	4999	9999	444%
R744	-51.6	1160	16.845	342	2.001	0.084	0.129	1	8.6E+05	2978	9999	444%
R729	-193.3	834	3.495	228	2.742	0.042	0.130	1	1.2E+06	2466	9999	444%
R740	-184.3	1386	6.519	160	1.149	0.057	0.097	1	1.5E+06	2734	9999	444%
R702	-254.2	72	0.894	454	8.067	0.010	0.265	0	4.5E+05	907	9607	427%
R728	-205.0	846	1.433	210	2.036	0.062	0.143	1	8.5E+05	1378	7902	351%
R114	12.9	1491	10.560	133	0.997	0.098	0.041	2	9.5E+05	3372	5554	247%
R14	-146.1	1696	1.945	143	0.857	0.111	0.058	2	9.6E+05	1755	4062	180%
R717	-72.7	728	0.092	1474	4.131	0.352	0.426	3	1.3E+05	222	3782	168%
R50	-177.5	445	0.429	538	3.420	0.208	0.108	7	1.3E+05	525	2277	101%
R113	-16.4	1660	0.562	164	0.890	0.285	0.050	5	3.6E+05	843	1698	75%
R704	-270.7	145	1.866	23	2.274	1.233	0.467	6	7.4E+03	84	1569	70%
R718	5.0	1000	0.007	2478	4.181	1.520	0.572	11	4.1E+04	58	1329	59%
R115	-94.4	1729	0.356	144	0.856	0.338	0.039	7	3.2E+05	767	1187	53%
R31	-152.8	1826	0.000	898	9.990	0.010	0.071	1	1.1E+07	405	1146	51%
R502	-101.1	1546	0.228	194	0.969	0.508	0.051	10	1.9E+05	490	1009	45%
R23	-150.0	2100	0.010	2812	9.990	3.341	0.129	258	3.9E+04	175	907	40%
R134a	-98.3	1578	0.047	261	1.187	0.536	0.075	8	1.8E+05	290	871	39%
R1150	-146.1	625	0.085	539	2.400	0.630	0.100	15	6.2E+04	217	870	39%
R503	-146.1	1621	0.048	212	1.090	0.922	0.079	13	1.1E+05	217	688	31%
R21	-68.2	1571	0.080	274	0.898	0.815	0.060	12	1.2E+05	270	647	29%
R40	-92.7	1106	0.052	463	1.195	0.978	0.092	13	7.1E+04	169	619	27%
R732	-213.8	1285	0.040	239	1.688	2.291	0.134	29	3.5E+04	113	608	27%
R504	-138.6	1833	0.011	252	1.428	1.470	0.093	23	7.8E+04	128	478	21%
R152a	-113.6	1184	0.006	402	1.541	1.415	0.090	24	5.2E+04	91	325	14%
R505	-118.2	1529	0.023	207	0.610	4.491	0.073	38	2.1E+04	66	193	9%
R506	-109.4	1591	0.004	313	1.335	4.690	0.078	80	2.1E+04	56	175	8%
R142b	-125.4	1440	0.001	284	1.123	8.251	0.066	141	1.1E+04	24	63	3%
R1270	-146.1	723	0.001	539	2.092	18.128	0.095	399	2.5E+03	12	46	2%
R11	-105.5	1759	0.001	226	0.784	14.134	0.044	249	7.8E+03	25	45	2%
R500	-140.0	1562	0.001	250	0.844	14.711	0.049	254	6.6E+03	21	41	2%
R22	-152.4	1708	0.000	300	1.068	46.675	0.059	852	2.3E+03	7	17	1%
R600	-133.3	730	0.000	492	2.01	54.367	0.102	1071	8.4E+02	4	16	1%
R170	-177.8	646	0.000	589	2.36	99.999	0.091	2606	4.0E+02	4	13	1%
R13b1	-162.8	2299	0.000	186	1.784	99.999	0.030	6014	1.4E+03	9	11	0%
R13	-176.1	1848	0.000	192	0.800	99.999	0.039	2069	1.2E+03	6	10	0%
R12	-152.1	1816	0.000	215	0.864	99.999	0.040	2175	1.1E+03	5	8	0%
R600a	-154.4	736	0.000	477	1.745	99.999	0.087	2009	4.6E+02	1	5	0%
R290	-182.6	728	0.000	558	1.938	99.999	0.079	2440	4.5E+02	1	3	0%

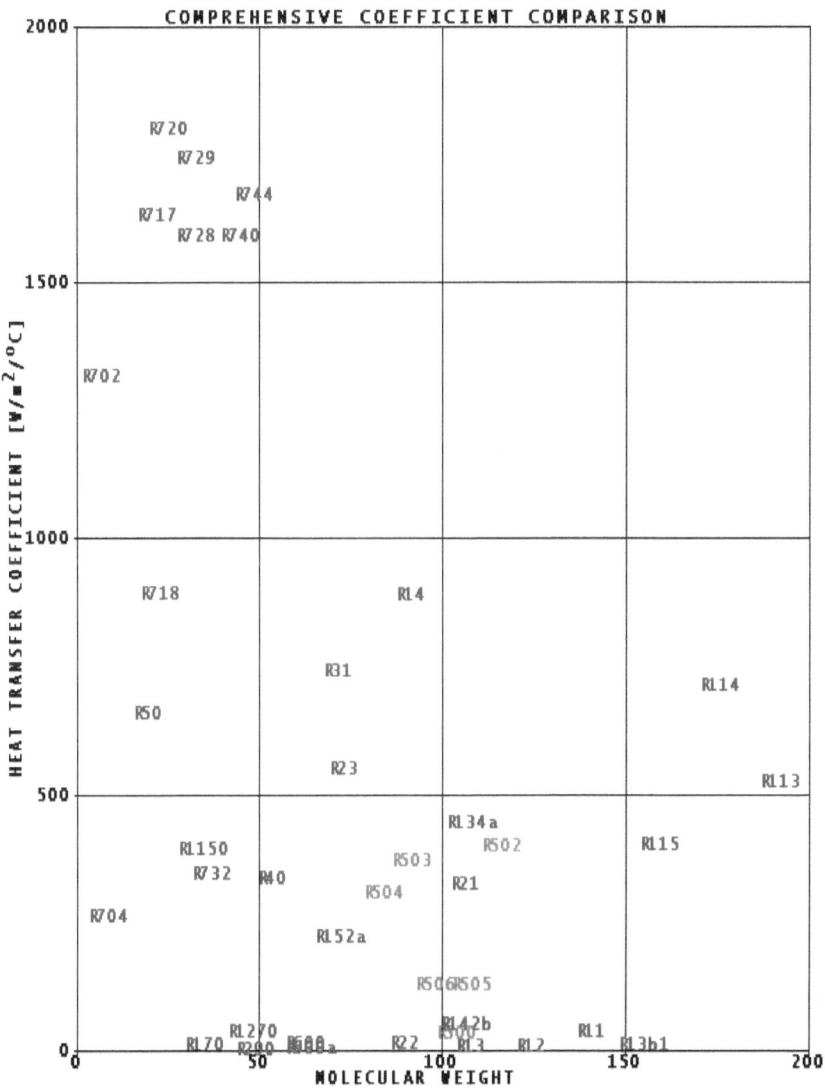

COMPREHENSIVE COEFFICIENT COMPARISON

Chapter 10. Conclusions & Recommendations

In this text we have considered thirty-nine common refrigerants, two cycles (vapor compression refrigeration and vapor expansion power), and four heat transfer coefficients (forced convective cooling and heating, condensation, and forced convective boiling). The six common azeotropes (500 series) have shown themselves to be unremarkable in these comparisons, while the 700 series (low molecular weight, nonflammable) ones have stood out. Azeotropes do exhibit unusual behavior, even if the thermodynamic and transport properties of these six have not resulted in superior performance in either of the two cycles considered, there are seventy-seven more refrigerant azeotropes that have been identified. These are listed in the following table:

ASHRAE number	IUPAC name	molecular formula	MW au	Tb °K	Tc °K	Pc kPa
R400a	R12/114(50/50 wt%)	50%/CCl2F2/50%/C2F4Cl2	141.6			
R400b	R12/114(60/40 wt%)	60%/CCl2F2/40%/C2F4Cl2	136.9			
R401A	R22/152a/124 (53/13/34)	53%/CHClF2/13%/C2H4F2/34%/C2HF4Cl	94.4	238.8	378.4	4613
R401B	R22/152a/124 (61/11/28)	61%/CHClF2/11%/C2H4F2/28%/C2HF4Cl	92.8	237.5	376.7	4682
R401C	R22/152a/124 (33/15/52)	33%/CHClF2/15%/C2H4F2/52%/C2HF4Cl	101.0	242.7		
R402A	R125/290/22 (60/2/38)	60%/C2HF5/2%/C3H8/38%//CHClF2	101.5	224.0	349.2	4234
R402B	R125/290/22 (38/2/60)	38%/C2HF5/2%/C3H8/60%/CHClF2	94.7	226.0	356.2	4525
R403A	R290/22/218 (5/75/20)	5%/C3H8/75%/CHClF2/20%/C3F8	92.0	229.2	364.4	4688
R403B	R290/22/218 (5/56/39)	5%/C3H8/56%/CHClF2/39%/C3F8	103.3	229.4	361.9	4399
R404A	R125/143a/134a (44/52/4)	44%/C2HF5/52%/C2H3F3/4%/C2H2F4	97.6	226.6	345.3	3735
R405A	R22/152a/142b/C318 (45/7/5.5/42.5)	45%/CHClF2/7%/C2H4F2/5.5%/C2H3F2Cl/42.5%/C4F8	111.9	240.3	379.2	4289
R406A	R22/600a/142b (55/4/41)	55%/CHClF2/4%/C4H10/41%/C2H3F2Cl	89.9	240.5	389.7	4881
R406B	R22/600a/142b (65/4/31)	65%/CHClF2/4%/C4H10/31%/C2H3F2Cl	88.6			
R407A	R32/125/134a (20/40/40)	20%/CH2F2/40%/C2HF5/40%/C2H2F4	90.1	228.0	355.1	4487

R407B	R32/125/134a (10/70/20)	10%/CH2F2/70%/C2HF5/20%/C2H2F4	102.9	226.4	347.5	4083
R407C	R32/125/134a (23/25/52)	23%/CH2F2/25%/C2HF5/52%/C2H2F4	86.2	229.4	359.2	4634
R407D	R32/125/134a (15/15/70)	15%/CH2F2/15%/C2HF5/70%/C2H2F4	91.0	233.8	364.7	4483
R407E	R32/125/134a (25/15/60)	25%/CH2F2/15%/C2HF5/60%/C2H2F4	83.8	230.4	361.9	4734
R407F	R32/125/134a (30/30/40)	30%/CH2F2/30%/C2HF5/40%/C2H2F4	82.1	227.1		
R408A	R125/143a/22 (7/46/47)	7%/C2HF5/46%/C2H3F3/47%/CHCIF2	87.0	227.7	356.7	4340
R409A	R22/124/142b (60/25/15)	60%/CHCIF2/25%/C2HF4CI/15%/C2H3F2CI	97.4	237.8	380.1	4600
R409B	R22/124/142b (65/25/10)	65%/CHCIF2/25%/C2HF4CI/10%/C2H3F2CI	96.7	236.7	377.5	4711
R410A	R32/125 (50/50)	50%/CH2F2/50%/C2HF5	72.6	221.6	343.3	4770
R410B	R32/125 (45/55)	45%/CH2F2/55%/C2HF5	75.6	221.7		
R411A	R1270/22/152a (1.5/87.5/11)	1.5%/C3H6/87.5%/CHCIF2/11%/C2H4F2	82.4	233.5	372.2	4954
R411B	R1270/22/152a (3/94/3)	3%/C3H6/94%/CHCIF2/3%/C2H4F2	83.1	231.6	369.1	4947
R411C	R1270/22/152a (3/95.5/1.5)	3%/C3H6/95.5%/CHCIF2/1.5%/C2H4F2	83.4		368.7	4950
R412A	R22/218/142b (70/5/25)	70%/CHCIF2/5%/C3F8/25%/C2H3F2CI	92.2	236.8	380.7	4881
R413A	R218/134a/600a (9/88/3)	9%/C3F8/88%/C2H2F4/3%/C4H10	104.0	243.9	374.5	4240
R414A	R22/124/600a/142b (51/28.5/4/16.5)	51%/CHCIF2/28.5%/C2HF4CI/4%/C4H10/16.5%/C2H3F2CI	96.9	239.2	383.9	4702
R414B	R22/124/600a/142b (50/39/1.5/9.5)	50%/CHCIF2/39%/C2HF4CI/1.5%/C4H10/9.5%/C2H3F2CI	101.5	238.8	381.2	4592
R415A	R22/152a (82/18)	82%/CHCIF2/18%/C2H4F2	81.9	235.7		
R415B	R22/152a (25/75)	25%/CHCIF2/75%/C2H4F2	70.2	249.8		
R416A	R134a/124/600 (59/39.5/1.5)	59%/C2H2F4/39.5%/C2HF4CI/1.5%/C4H10	111.9	235.2	381.4	4020
R417A	R125/134a/600 (46.6/50/3.4)	46.6%/C2HF5/50%/C2H2F4/3.4%/C4H10	106.7	232.0	363.0	4102

R417B	R125/134a/600 (79/18.3/2.7)	79%/C2HF5/18.3%/C2H2F4/2.7%/C4H10	113.1	228.3		
R418A	R290/22/152a (1.5/96/2.5)	1.5%/C3H8/96%/CHClF2/2.5%/C2H4F2	84.6	230.6		
R419A	R125/134a/E170 (77/19/4)	77%/C2HF5/19%/C2H2F4/4%/C2H6O	109.3	248.2		
R420A	R134a/142b (88/12)	88%/C2H2F4/12%/C2H3F2Cl	101.8	238.8		
R421A	R125/134a (58/42)	58%/C2HF5/42%/C2H2F4	111.7			
R421B	R125/134a (85/15)	85%/C2HF5/15%/C2H2F4	116.9			
R422A	R125/134a/600a (85.1/11.5/3.4)	85.1%/C2HF5/11.5%/C2H2F4/3.4%/C4H10	113.6			
R422B	R125/134a/600a (55/42/3)	55%/C2HF5/42%/C2H2F4/3%/C4H10	108.5			
R422C	R125/134a/600a (82/15/3)	82%/C2HF5/15%/C2H2F4/3%/C4H10	113.4			
R422D	R125/134a/600a (65.1/31.5/3.4)	65.1%/C2HF5/31.5%/C2H2F4/3.4%/C4H10	109.9			
R423A	R134a/227ea (52.5/47.5)	52.5%/C2H2F4/47.5%/C3HF7	126.0			
R424A	R125/134a/600a/600/601a (50.5/47/0.9/1/0.6)	50.5%/C2HF5/47%/C2H2F4/1.9%/C4H10/0.6%/C5H12	108.4			
R425A	R32/134a/227ea (18.5/69.5/12)	18.5%/CH2F2/69.5%/C2H2F4/12%/C3HF7	90.3			
R426A	R125/134a/600/601a (5.1/93/1.3/0.6)	5.1%/C2HF5/93%/C2H2F4/1.3%/C4H10/0.6%/C5H12	101.6			
R427A	R32/125/143a/134a (15/25/10/50)	15%/CH2F2/25%/C2HF5/10%/C2H3F3/50%/C2H2F4	90.4			
R428A	R125/143a/290/600a (77.5/20/0.6/1.9)	77.5%/C2HF5/20%/C2H3F3/0.6%/C3H8/1.9%/C4H10	107.5			
R429A	RE170/152a/600a (60/10/30)	60%/C2H6O/10%/C2H4F2/30%/C4H10	50.8			
R430A	R152a/600a (76/24)	76%/C2H4F2/24%/C4H10	64.0			
R431A	R290/152a (71/29)	71%/C3H8/29%/C2H4F2	48.8			
R432A	R1270/E170 (80/20)	80%/C3H6/20%/C2H6O	42.8			
R433A	R1270/290 (30/70)	30%/C3H6/70%/C3H8	43.5	228.6		

R433B	R1270/290 (5/95)	5%/C3H6/95%/C3H8	44.0	230.5		
R433C	R1270/290 (25/75)	25%/C3H6/75%/C3H8	43.6	228.9		
R434A	R125/143a/134a/600a (63.2/18/16/2.8)	63.2%/C2HF5/18%/C2H3F3/16%/C2H2F4/2.8%/C4H10	105.7	228.2		
R435A	RE170/152a (80/20)	80%/C2H6O/20%/C2H4F2	49.0	247.1		
R436A	R290/600a (56/44)	56%/C3H8/44%/C4H10	49.3	238.9		
R436B	R290/600a (52/48)	52%/C3H8/48%/C4H10	49.9	239.8		
R437A	R125/134a/600/601 (19.5/78.5/1.4/0.6)	19.5%/C2HF5/78.5%/C2H2F4/1.4%/C4H10/0.6%/C5H12	103.7	240.3		
R438A	R32/125/134a/600/601a (8.5/45/44.2/1.7/0.6)	8.5%/CH2F2/45%/C2HF5/44.2%/C2H2F4/1.7%/C4H10/0.6%/C5H12	99.1	230.2		
R439A	R32/125/600a (50/47/3)	50%/CH2F2/47%/C2HF5/3%/C4H10	71.2	221.2		
R440A	R290/134a/152a (.6/1.6/97.8)	.6%/C3H8/1.6%/C2H2F4/97.8%/C2H4F2	66.2	247.7		
R441A	R170/290/600a/600 (3.1/54.8/6/36.1)	3.1%/C2H6/54.8%/C3H8/42.1%/C4H10	48.3	231.3		
R455A	R1234yf/32/744 (75.5%/21.5%/3%)	75.5% C3H2F4 /21.5% CH2F2 /3% CO2			358.8	4660
R458A	R32/125/134a/227ea/236fa (20.5%/4.0%/61.4%/13.5%/0.6%)	20.5% C2F2/4.0% C2HF5/61.4% C2H3F3/13.5% C3HF7/0.6% C3H2F6				
R500	R12/152a (73.8/26.2)	73.8%/CCl2F2/26.2%/C2H4F2	99.3	240.2	375.3	4173
R501	R22/12 (75/25)	75%/CHClF2/25%/CCl2F2	93.1	232.2	369.3	4764
R502	R22/115 (48.8/51.2)	48.8%/CHClF2/51.2%/C2F5Cl	111.6	228.2	353.9	4019
R503	R23/13 (40.1/59.9)	40.1%/CHF3/59.9%/CClF3	87.2	185.2	291.6	4265
R504	R32/115 (48.2/51.8)	48.2%/CH2F2/51.8%/C2F5Cl	79.2	216.2	335.3	4439
R505	R12/31 (78/22)	78%/CCl2F2/22%/CH2FCl	103.5	243.2	390.9	4730
R506	R31/114 (55.1/44.9)	55.1%/CH2FCl/44.9%/C2F4Cl2	93.7	261.2	415.4	5157
R507	R125/143a (50/50)	50%/C2HF5/50%/C2H3F3	98.9	226.5	343.9	3715
R508	R23/116 (39/61)	39%/CHF3/61%/C2F6	100.1	187.2	284.2	3701
R508B	R23/116 (46/54)	46%/CHF3/54%/C2F6	95.4	184.9	285.2	3834

R509	R22/218 (44/56)	44%/CHClF2/56%/C3F8	124.0	226.2	360.4	4033
R510	RE170/600a (88/12)	88%/C2H6O/12%/C4H10	47.2	248.0		
R511	R290/E170 (95/5)	95%/C3H8/5%/C2H6O	44.2	231.1		
R513A	R1234yf/134a (56%/44%)	56% C3H2F4 /44% C2H2F2	108.4	243.8	369.7	3766

We have not investigated all of these here, simply because the work has not yet been done to complete the thermodynamic and transport properties and build these into software such as the Excel AddIn. Even NIST's REFPROP doesn't contain all of these substances. We have shown how to extend this work, for instance with the BWRS EoS (Appendix D) and the work of Prausnitz (Chapter 8). Carrying this work forward would make an excellent dissertation topic. Please let me know if you plan to pursue this valuable effort and how it turns out.

Appendix A. Critical Constants

All chemically stable fluids exhibit a vapor dome (locus of points corresponding to liquid and vapor states in equilibrium) and a critical point (where the saturated liquid and vapor are indistinguishable). Properties at the critical point (pressure, temperature, and specific volume or density) are crucial to developing and utilizing thermodynamic properties. The following list of critical constants for common substances (listed by increasing molecular weight) has been assembled from a variety of sources and converted to uniform units.

name	formula	MW	Tc	Pc	Vc	R	Zc	Pitzer
			°K	kPa	m³/kg	J/kg/K	-	acentric
Hydrogen	H2	2.02	33.2	1297	0.032200	4.124	0.305	-0.2216
Helium	He	4.00	5.2	227	0.014296	2.077	0.301	0.0000
Methane	CH4	16.04	190.9	4599	0.006143	0.518	0.286	0.0108
Ammonia	NH3	17.03	405.5	11356	0.004251	0.488	0.244	0.2558
Water Vapor	H2O	18.02	647.1	22064	0.003106	0.462	0.229	0.3445
Ethyne (acetylene)	C2H2	26.04	308.3	6139	0.004326	0.319	0.270	0.1949
Carbon Monoxide	CO	28.01	132.9	3494	0.003627	0.297	0.321	0.0477
Nitrogen	N2	28.01	126.2	3396	0.003203	0.297	0.290	0.0370
Ethene (ethylene)	C2H4	28.05	282.3	5040	0.004657	0.296	0.281	0.0860
Air	N2+O2	28.96	132.4	3771	0.003228	0.287	0.320	0.0000
Ethane	C2H6	30.07	305.3	4872	0.004838	0.277	0.279	0.0972
Oxygen	O2	32.00	154.6	5043	0.002291	0.260	0.288	0.0216
Hydrogen Sulfide	H2S	34.08	373.4	8963	0.002878	0.244	0.283	0.0948
Argon	Ar	39.95	150.8	4870	0.001873	0.208	0.291	0.0010
Propene (propylene)	C3H6	42.08	365.5	4665	0.004476	0.198	0.289	0.1404
Cyclopropane	C3H6	42.08	397.8	5490	0.004008	0.198	0.280	0.1300
Carbon Dioxide	CO2	44.01	304.1	7377	0.002135	0.189	0.274	0.2667
n-Propane	C3H8	44.10	369.8	4244	0.004539	0.189	0.276	0.1515
Iso-Butene (2-Methyl Propene)	C4H8	56.11	417.9	4000	0.004251	0.148	0.275	0.1953
1-Butene (butylene)	C4H8	56.11	419.9	4043	0.004264	0.148	0.277	0.1914
trans-Butene	C4H8	56.11	428.6	3964	0.004239	0.148	0.265	0.2034
Butenes (average)	C4H8	56.11	432.4	4248	0.004120	0.148	0.273	0.1953
cis-Butene	C4H8	56.11	435.5	4243	0.004164	0.148	0.274	0.2054
Cyclobutane	C4H8	56.11	460.0	4990	0.004133	0.148	0.303	0.1810
i-Butane	C4H10	58.12	407.8	3640	0.004457	0.143	0.278	0.1852
n-Butane	C4H10	58.12	425.1	3798	0.004389	0.143	0.274	0.1981
Butanes (average)	C4H10	58.12	416.5	3712	0.004426	0.143	0.276	0.1917
Sulfur Dioxide	SO2	64.07	430.8	7881	0.001904	0.130	0.268	0.2548
1-Pentene	C5H10	70.13	464.7	3513	0.004208	0.119	0.268	0.2313
Pentenes (average)	C5H10	70.13	488.1	4011	0.003958	0.119	0.274	0.2126
Cyclopentane	C5H10	70.13	511.5	4508	0.003708	0.119	0.276	0.1938
i-Pentane	C5H12	72.15	460.4	3381	0.004239	0.115	0.270	0.2286

name	formula	MW	Tc	Pc	Vc	R	Zc	Pitzer
			°K	kPa	m³/kg	J/kg/K	-	acentric
1-Pentene	C5H10	70.13	464.7	3513	0.004208	0.119	0.268	0.2313
Pentenes (average)	C5H10	70.13	488.1	4011	0.003958	0.119	0.274	0.2126
Cyclopentane	C5H10	70.13	511.5	4508	0.003708	0.119	0.276	0.1938
i-Pentane	C5H12	72.15	460.4	3381	0.004239	0.115	0.270	0.2286
n-Pentane	C5H12	72.15	469.7	3370	0.004308	0.115	0.268	0.2510
Neopentane (2,2-Dimethyl Propane)	C5H12	72.15	433.7	3199	0.004201	0.115	0.269	0.1965
Pentanes (average)	C5H12	72.15	454.6	3315	0.004270	0.115	0.270	0.2254
Benzene	C6H6	78.11	562.1	4898	0.003315	0.106	0.271	0.2090
Cyclohexane	C6H12	84.16	553.5	4073	0.003658	0.099	0.272	0.2094
n-Hexane	C6H14	86.18	507.5	3012	0.004270	0.096	0.263	0.2990
2,2-Dimethyl Butane (Neohexane)	C6H14	86.18	488.7	3080	0.004164	0.096	0.272	0.2334
2-Methylpentane	C6H14	86.18	497.5	3010	0.004258	0.096	0.267	0.2780
Hexanes (average)	C6H14	86.18	499.2	3060	0.004245	0.096	0.270	0.2710
2,3-Dimethyl Butane	C6H14	86.18	499.9	3130	0.004151	0.096	0.269	0.2481
3-Methylpentane	C6H14	86.18	504.4	3120	0.004258	0.096	0.273	0.2736
n-Heptane	C7H16	100.20	540.3	2736	0.004270	0.083	0.261	0.3483
n-Octane	C8H18	114.23	568.8	2487	0.004308	0.073	0.259	0.3978
n-Nonane	C9H20	128.26	594.7	2280	0.004326	0.065	0.256	0.4425
n-Decane	C10H22	142.28	617.7	2100	0.004389	0.058	0.255	0.4881

The following table lists common refrigerants from several sources.

Name	Formula	MW	Tf	Tc	Pc	Vc	R	Zc	Pitzer
		au	°K	°K	kPa	m³/kg	J/kg/°K	-	acentric
R11	CCl3F	137.37	162.0	471.2	4409	0.00181	0.0605	0.2792	0.1915
R12	CF2Cl2	120.93	115.4	385.2	4115	0.00179	0.0688	0.2785	0.1764
R13	CClF3	104.46	122.0	301.5	3870	0.00173	0.0796	0.2790	0.1792
R13b1	CBrF3	148.91	105.4	340.2	3964	0.00134	0.0558	0.2802	0.1722
R14	CF4	88.01	89.3	227.7	3745	0.00160	0.0945	0.2782	0.1740
R21	CHCl2F	102.92	200.0	451.6	5167	0.00185	0.0808	0.2616	0.2038
R22	CHClF2	86.48	113.2	369.2	4977	0.00191	0.0961	0.2672	0.2220
R23	CHF3	70.01	118.2	299.1	4836	0.00190	0.1188	0.2593	0.2654
R40	CH3Cl	50.49	175.4	416.3	6759	0.00279	0.1647	0.2754	0.1634
R50	CH4	16.04	90.9	190.7	4641	0.00618	0.5184	0.2903	0.0112
R113	Cl2FC-CClF2	187.38	238.2	487.2	3392	0.00179	0.0444	0.2802	0.2518
R114	ClF2C-CClF2	170.92	179.3	418.8	3257	0.00172	0.0486	0.2756	0.2544
R115	F3C-CClF2	154.47	167.0	353.1	3129	0.00163	0.0538	0.2678	0.2494
R134a	CH2FCF3	102.03	176.5	374.2	4059	0.00195	0.0815	0.2600	0.3270
R142b	CH3CClF2	100.50	142.0	410.3	4246	0.00230	0.0827	0.2876	0.2501
R152a	CHF2CH3	66.05	156.2	386.4	4520	0.00272	0.1259	0.2525	0.2694

Name	Formula	MW	Tf	Tc	Pc	Vc	R	Zc	Pitzer
		au	°K	°K	kPa	m³/kg	J/kg/°K	-	acentric
R170	C2H6	30.07	90.4	305.4	4894	0.00518	0.2765	0.3003	0.0988
R290	C3H8	44.10	85.5	369.9	4251	0.00454	0.1886	0.2764	0.1523
R500	R12/R152a	99.30	128.2	375.3	4173	0.00201	0.0837	0.2672	0.2170
R502	R22/R115	111.60	167.0	355.3	4075	0.00178	0.0745	0.2746	0.2186
R503	R23/R13	87.20	122.0	292.7	4359	0.00177	0.0953	0.2770	0.1978
R504	R32/R115	79.20	129.6	335.3	4439	0.00182	0.1050	0.2293	0.2650
R505	R12/R31	103.50	150.0	390.9	4727	0.00186	0.0803	0.2804	0.1776
R506	R31/R114	93.70	158.7	414.8	5167	0.00181	0.0887	0.2547	0.2345
R600	C4H10	58.10	135.0	425.2	3797	0.00438	0.1431	0.2735	0.2015
R600a	C4H10	58.10	113.4	408.1	3648	0.00452	0.1431	0.2823	0.1848
R702	H2	2.02	13.8	33.0	1293	0.03182	4.1242	0.3024	-0.2200
R704	He	4.00	2.0	5.2	248	0.01436	2.0776	0.3297	-0.3900
R717	NH3	17.03	195.4	405.4	11330	0.00425	0.4882	0.2434	0.2530
R718	H2O	18.02	273.2	647.3	22088	0.00315	0.4615	0.2333	0.3340
R720	Ne	20.18	24.5	44.4	2654	0.00207	0.4120	0.3003	0.0000
R728	N2	28.01	63.2	126.2	3394	0.00321	0.2968	0.2913	0.0400
R729	Air	28.97	56.9	132.4	3774	0.00305	0.2870	0.3026	0.0225
R732	O2	32.00	54.4	154.8	5080	0.00229	0.2598	0.2899	0.0229
R740	Ar	39.95	83.8	150.9	4898	0.00187	0.2081	0.2912	-0.0039
R744	CO2	44.01	216.5	304.2	7392	0.00216	0.1889	0.2773	0.2280
R1150	CH2=CH2	28.05	104.0	283.1	5117	0.00437	0.2964	0.2664	0.0787
R1270	CH3CH=CH2	42.10	88.0	364.9	4621	0.00454	0.1975	0.2911	0.1445

These tables can be found in the online archive in folder examples in spreadsheet critical_data.xls. As it is in this form, you can sort the lists on other columns to facilitate comparisons. The figure below can be found in critical.lay and critical.plt.

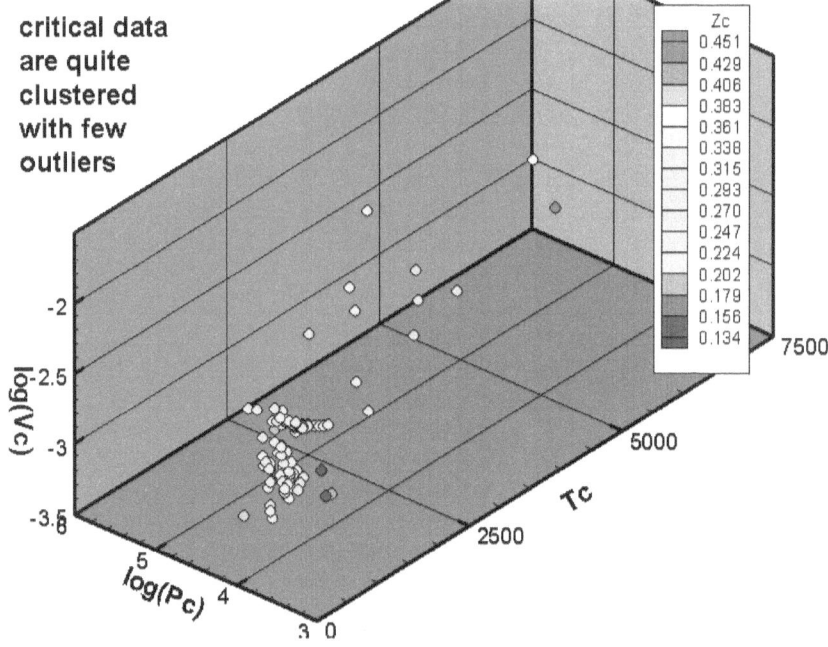

Appendix B. Chlorofluoromethane (R31)

R31 has been known for many decades, but has fallen out of use and disappeared from the market due to safety factors and association with global warming. Still, it is important to understand from a thermodynamic perspective and also because it is one of two components (along with R114) that comprise azeotrope R506. The properties can no longer be found on the Web. I have collected the existing information and pieced together a consistent set of thermodynamic properties based on critical data, normal boiling point data, zero density specific heat data, Pitzer acentric factor, liquid density, and isothermal compressibility. These scattered data points, along with dimensionless curves based on similar fluids, have been combined to produce the following table.

Thermodynamic Properties of Chlorofluoromethane (R31)

Name	Formula	MW	Tc	Pc	Vc	R	Zc
R31	CH2FCl	68.478	424.9	5131	0.002315	0.1214	0.2302
Ts	Ps	Vf	Vg	Hf	Hg	Sf	Sg
°K	kPa	m³/kg	m³/kg	kJ/kg	kJ/kg	kJ/kg/°K	kJ/kg/°K
115.4	0.00002813	0.0005886	497917	-231.0	252.0	-1.5516	2.6350
117.1	0.00004251	0.0005904	334390	-226.1	252.9	-1.4989	2.5925
118.8	0.00006344	0.0005922	227367	-221.4	253.8	-1.4489	2.5515
120.5	0.0000935	0.0005940	156439	-216.9	254.7	-1.4015	2.5117
122.2	0.0001363	0.0005958	108865	-212.5	255.5	-1.3566	2.4733
123.9	0.0001965	0.0005977	76585	-208.3	256.4	-1.3138	2.4361
125.6	0.0002802	0.0005995	54439	-204.2	257.3	-1.2731	2.4000
127.3	0.000396	0.0006013	39085	-200.2	258.2	-1.2343	2.3651
129.1	0.000553	0.0006032	28331	-196.3	259.1	-1.1972	2.3312
130.8	0.000766	0.0006050	20726	-192.5	259.9	-1.1618	2.2984
132.5	0.001052	0.0006069	15296	-188.9	260.8	-1.1279	2.2666
134.2	0.001431	0.0006087	11385	-185.3	261.7	-1.0955	2.2357
135.9	0.001931	0.0006106	8543	-181.8	262.5	-1.0643	2.2057
137.6	0.002586	0.0006125	6460	-178.4	263.4	-1.0345	2.1767
139.3	0.003436	0.0006144	4923	-175.1	264.3	-1.0057	2.1485
141.0	0.00453	0.0006163	3778	-171.9	265.2	-0.9781	2.1211
142.7	0.00593	0.0006182	2920	-168.7	266.0	-0.9515	2.0945
144.4	0.00772	0.0006202	2272	-165.6	266.9	-0.9258	2.0687
146.2	0.00997	0.0006221	1779	-162.6	267.8	-0.9010	2.0436
147.9	0.01280	0.0006240	1402	-159.6	268.7	-0.8771	2.0192
149.6	0.01633	0.0006260	1112	-156.7	269.5	-0.8540	1.9955
151.3	0.02072	0.0006280	886.6	-153.8	270.4	-0.8316	1.9725
153.0	0.0261	0.0006299	711.0	-151.0	271.3	-0.8100	1.9501
154.7	0.0328	0.0006319	573.3	-148.2	272.1	-0.7890	1.9283
156.4	0.0409	0.0006339	464.7	-145.5	273.0	-0.7686	1.9071

Thermodynamic Properties of Chlorofluoromethane (R31) contd.

Ts	Ps	Vf	Vg	Hf	Hg	Sf	Sg
°K	kPa	m³/kg	m³/kg	kJ/kg	kJ/kg	kJ/kg/°K	kJ/kg/°K
158.1	0.0507	0.0006359	378.5	-142.8	273.9	-0.7489	1.8865
159.8	0.0626	0.0006379	309.9	-140.2	274.8	-0.7297	1.8664
161.5	0.0770	0.0006400	254.9	-137.6	275.6	-0.7111	1.8469
163.3	0.0941	0.0006420	210.6	-135.0	276.5	-0.6929	1.8279
165.0	0.1146	0.0006441	174.8	-132.5	277.4	-0.6753	1.8094
166.7	0.1389	0.0006461	145.7	-130.0	278.3	-0.6581	1.7913
168.4	0.1677	0.0006482	121.9	-127.5	279.1	-0.6414	1.7738
170.1	0.202	0.0006503	102.4	-125.1	280.0	-0.6250	1.7566
171.8	0.241	0.0006524	86.42	-122.7	280.9	-0.6091	1.7400
173.5	0.288	0.0006545	73.18	-120.3	281.7	-0.5939	1.7233
175.2	0.342	0.0006566	62.19	-118.0	282.6	-0.5788	1.7074
176.9	0.405	0.0006587	53.04	-115.7	283.5	-0.5640	1.6919
178.6	0.478	0.0006609	45.40	-113.4	284.3	-0.5496	1.6768
180.4	0.561	0.0006630	38.98	-111.1	285.2	-0.5355	1.6620
182.1	0.658	0.0006652	33.58	-108.9	286.1	-0.5217	1.6476
183.8	0.768	0.0006674	29.02	-106.7	286.9	-0.5082	1.6335
185.5	0.895	0.0006695	25.15	-104.5	287.7	-0.4949	1.6197
187.2	1.038	0.0006717	21.86	-102.3	288.6	-0.4820	1.6062
188.9	1.202	0.0006740	19.06	-100.2	289.4	-0.4693	1.5931
190.6	1.387	0.0006762	16.67	-98.0	290.3	-0.4569	1.5802
192.3	1.596	0.0006784	14.61	-95.9	291.1	-0.4447	1.5676
194.0	1.83	0.0006807	12.84	-93.8	291.9	-0.4328	1.5553
195.7	2.10	0.0006830	11.31	-91.8	292.7	-0.4211	1.5433
197.5	2.39	0.0006852	9.995	-89.7	293.6	-0.4096	1.5315
199.2	2.73	0.0006875	8.850	-87.7	294.4	-0.3983	1.5199
200.9	3.10	0.0006898	7.855	-85.7	295.2	-0.3873	1.5086
202.6	3.51	0.0006922	6.988	-83.7	296.0	-0.3764	1.4976
204.3	3.97	0.0006945	6.230	-81.7	296.7	-0.3657	1.4868
206.0	4.48	0.0006969	5.567	-79.7	297.5	-0.3552	1.4761
207.7	5.04	0.0006992	4.984	-77.8	298.3	-0.3449	1.4657
209.4	5.66	0.0007016	4.472	-75.9	299.1	-0.3347	1.4555
211.1	6.34	0.0007040	4.020	-73.9	299.8	-0.3247	1.4455
212.9	7.10	0.0007064	3.621	-72.0	300.6	-0.3149	1.4357

84

Thermodynamic Properties of Chlorofluoromethane (R31) contd.

Ts	Ps	Vf	Vg	Hf	Hg	Sf	Sg
°K	kPa	m³/kg	m³/kg	kJ/kg	kJ/kg	kJ/kg/°K	kJ/kg/°K
214.6	7.92	0.0007089	3.267	-70.1	301.3	-0.3052	1.4261
216.3	8.83	0.0007113	2.954	-68.2	302.1	-0.2957	1.4167
218.0	9.82	0.0007138	2.675	-66.4	302.8	-0.2863	1.4074
219.7	10.91	0.0007163	2.427	-64.5	303.5	-0.2770	1.3984
221.4	12.09	0.0007188	2.205	-62.7	304.3	-0.2678	1.3895
223.1	13.37	0.0007213	2.008	-60.8	305.0	-0.2588	1.3807
224.8	14.77	0.0007238	1.830	-59.0	305.7	-0.2499	1.3722
226.5	16.29	0.0007264	1.671	-57.2	306.4	-0.2411	1.3638
228.2	17.93	0.0007289	1.528	-55.3	307.1	-0.2324	1.3555
230.0	19.71	0.0007315	1.400	-53.5	307.8	-0.2238	1.3475
231.7	21.63	0.0007341	1.284	-51.7	308.5	-0.2153	1.3396
233.4	23.70	0.0007367	1.179	-49.9	309.2	-0.2069	1.3318
235.1	25.93	0.0007394	1.085	-48.1	309.9	-0.1985	1.3242
236.8	28.33	0.0007420	0.9990	-46.3	310.6	-0.1903	1.3168
238.5	30.92	0.0007447	0.9212	-44.5	311.3	-0.1820	1.3095
240.2	33.69	0.0007474	0.8506	-42.7	312.0	-0.1739	1.3023
241.9	36.66	0.0007502	0.7864	-40.8	312.6	-0.1658	1.2953
243.6	39.84	0.0007529	0.7279	-39.0	313.3	-0.1578	1.2885
245.3	43.24	0.0007557	0.6745	-37.2	314.0	-0.1498	1.2818
247.1	46.87	0.0007584	0.6258	-35.4	314.7	-0.1418	1.2753
248.8	50.75	0.0007613	0.5812	-33.5	315.4	-0.1339	1.2689
250.5	54.88	0.0007641	0.5404	-31.7	316.1	-0.1260	1.2626
252.2	59.28	0.0007669	0.5030	-29.8	316.8	-0.1182	1.2565
253.9	63.96	0.0007698	0.4687	-28.0	317.5	-0.1103	1.2506
255.6	68.94	0.0007727	0.4372	-26.1	318.3	-0.1025	1.2448
257.3	74.22	0.0007756	0.4081	-24.2	319.0	-0.0947	1.2391
259.0	79.82	0.0007786	0.3814	-22.3	319.7	-0.0869	1.2336
260.7	85.75	0.0007816	0.3568	-20.4	320.5	-0.0791	1.2282
262.4	92.03	0.0007845	0.3340	-18.4	321.2	-0.0712	1.2230
264.2	98.67	0.0007876	0.3130	-16.5	322.0	-0.0634	1.2180
265.9	105.69	0.0007906	0.2936	-14.5	322.8	-0.0556	1.2131
267.6	113.10	0.0007937	0.2756	-12.5	323.6	-0.0477	1.2083
269.3	120.91	0.0007968	0.2589	-10.5	324.4	-0.0398	1.2037

Thermodynamic Properties of Chlorofluoromethane (R31) contd.

Ts	Ps	Vf	Vg	Hf	Hg	Sf	Sg
°K	kPa	m³/kg	m³/kg	kJ/kg	kJ/kg	kJ/kg/°K	kJ/kg/°K
271.0	129.15	0.0007999	0.2435	-8.4	325.2	-0.0319	1.1992
272.7	137.83	0.0008031	0.2291	-6.4	326.0	-0.0240	1.1949
274.4	146.96	0.0008063	0.2158	-4.3	326.9	-0.0160	1.1907
276.1	156.56	0.0008095	0.2033	-2.1	327.7	-0.0080	1.1866
277.8	166.65	0.0008127	0.1918	0.0	328.6	0.0000	1.1828
279.5	177.24	0.0008160	0.1810	2.2	329.5	0.0081	1.1790
281.3	188.35	0.0008193	0.1710	4.4	330.4	0.0162	1.1754
283.0	200.00	0.0008227	0.1616	6.6	331.3	0.0244	1.1719
284.7	212.21	0.0008261	0.1528	8.9	332.3	0.0326	1.1686
286.4	225.00	0.0008295	0.1446	11.2	333.3	0.0408	1.1654
288.1	238.37	0.0008329	0.1369	13.5	334.2	0.0491	1.1624
289.8	252.36	0.0008364	0.1298	15.9	335.2	0.0575	1.1595
291.5	266.98	0.0008399	0.1230	18.3	336.3	0.0659	1.1567
293.2	282.24	0.0008435	0.1167	20.7	337.3	0.0744	1.1540
294.9	298.18	0.0008470	0.1108	23.2	338.4	0.0829	1.1515
296.6	314.80	0.0008507	0.1052	25.7	339.4	0.0915	1.1491
298.4	332.13	0.0008543	0.09999	28.2	340.5	0.1001	1.1469
300.1	350.18	0.0008581	0.09507	30.8	341.6	0.1088	1.1447
301.8	368.99	0.0008618	0.09044	33.4	342.8	0.1176	1.1427
303.5	388.56	0.0008656	0.08608	36.0	343.9	0.1264	1.1408
305.2	408.92	0.0008694	0.08197	38.7	345.1	0.1352	1.1390
306.9	430.08	0.0008733	0.07809	41.4	346.2	0.1441	1.1373
308.6	452.08	0.0008772	0.07443	44.2	347.4	0.1531	1.1357
310.3	474.93	0.0008812	0.07097	47.0	348.6	0.1622	1.1343
312.0	498.64	0.0008852	0.06771	49.8	349.9	0.1713	1.1329
313.8	523.25	0.0008893	0.06462	52.7	351.1	0.1804	1.1316
315.5	548.78	0.0008934	0.06170	55.5	352.3	0.1896	1.1304
317.2	575.24	0.0008976	0.05894	58.5	353.6	0.1989	1.1293
318.9	602.66	0.0009018	0.05632	61.4	354.9	0.2082	1.1283
320.6	631.05	0.0009061	0.05383	64.4	356.1	0.2175	1.1274
322.3	660.45	0.0009105	0.05148	67.5	357.4	0.2269	1.1265
324.0	690.87	0.0009149	0.04925	70.5	358.7	0.2363	1.1258
325.7	722.34	0.0009193	0.04713	73.6	360.0	0.2458	1.1250

Thermodynamic Properties of Chlorofluoromethane (R31) contd.

Ts	Ps	Vf	Vg	Hf	Hg	Sf	Sg
°K	kPa	m³/kg	m³/kg	kJ/kg	kJ/kg	kJ/kg/°K	kJ/kg/°K
327.4	754.88	0.0009239	0.04512	76.8	361.3	0.2554	1.1244
329.1	788.51	0.0009285	0.04320	79.9	362.6	0.2649	1.1238
330.9	823.25	0.0009331	0.04139	83.1	363.9	0.2745	1.1232
332.6	859.13	0.0009379	0.03966	86.4	365.2	0.2842	1.1227
334.3	896.17	0.0009427	0.03801	89.6	366.6	0.2938	1.1223
336.0	934.4	0.0009475	0.03644	92.9	367.9	0.3035	1.1219
337.7	973.8	0.0009525	0.03495	96.2	369.1	0.3132	1.1215
339.4	1014.5	0.0009575	0.03353	99.5	370.4	0.3230	1.1211
341.1	1056.4	0.0009626	0.03217	102.9	371.7	0.3328	1.1208
342.8	1099.6	0.0009678	0.03087	106.3	373.0	0.3425	1.1204
344.5	1144.1	0.0009731	0.02964	109.7	374.2	0.3523	1.1201
346.2	1189.9	0.0009785	0.02846	113.2	375.5	0.3622	1.1198
348.0	1237.0	0.0009840	0.02733	116.6	376.7	0.3720	1.1195
349.7	1285.5	0.0009896	0.02625	120.1	377.9	0.3818	1.1191
351.4	1335.5	0.0009952	0.02522	123.6	379.1	0.3917	1.1188
353.1	1386.8	0.001001	0.02423	127.1	380.2	0.4015	1.1184
354.8	1439.6	0.001007	0.02329	130.6	381.4	0.4113	1.1180
356.5	1493.8	0.001013	0.02238	134.2	382.5	0.4211	1.1175
358.2	1549.5	0.001019	0.02152	137.8	383.5	0.4310	1.1170
359.9	1606.7	0.001025	0.02069	141.3	384.5	0.4407	1.1164
361.6	1665.5	0.001032	0.01989	144.9	385.5	0.4505	1.1158
363.3	1725.8	0.001038	0.01913	148.5	386.5	0.4603	1.1151
365.1	1787.8	0.001045	0.01839	152.1	387.4	0.4700	1.1143
366.8	1851.3	0.001052	0.01769	155.8	388.2	0.4797	1.1135
368.5	1916.4	0.001059	0.01701	159.4	389.0	0.4894	1.1125
370.2	1983.2	0.001066	0.01636	163.0	389.7	0.4990	1.1115
371.9	2051.7	0.001074	0.01573	166.6	390.4	0.5086	1.1103
373.6	2121.9	0.001081	0.01513	170.3	391.0	0.5182	1.1090
375.3	2193.8	0.001089	0.01455	173.9	391.5	0.5277	1.1075
377.0	2267.4	0.001097	0.01399	177.5	392.0	0.5372	1.1059
378.7	2342.8	0.001105	0.01345	181.2	392.4	0.5466	1.1042
380.4	2420.1	0.001114	0.01294	184.8	392.7	0.5560	1.1023
382.2	2499.1	0.001122	0.01243	188.4	392.9	0.5653	1.1002

Thermodynamic Properties of Chlorofluoromethane (R31) contd.

Ts	Ps	Vf	Vg	Hf	Hg	Sf	Sg
°K	kPa	m³/kg	m³/kg	kJ/kg	kJ/kg	kJ/kg/°K	kJ/kg/°K
383.9	2580.0	0.001131	0.01195	192.1	393.0	0.5745	1.0979
385.6	2662.7	0.001141	0.01148	195.7	393.0	0.5837	1.0954
387.3	2747.3	0.001151	0.01103	199.3	392.9	0.5929	1.0927
389.0	2833.9	0.001161	0.01059	202.9	392.7	0.6019	1.0898
390.7	2922.3	0.001171	0.01017	206.5	392.4	0.6110	1.0866
392.4	3012.7	0.001182	0.00976	210.1	391.9	0.6200	1.0832
394.1	3105.1	0.001193	0.00936	213.7	391.3	0.6289	1.0795
395.8	3199.5	0.001205	0.00897	217.4	390.6	0.6378	1.0755
397.5	3296.0	0.001218	0.00859	221.0	389.7	0.6467	1.0712
399.3	3394.4	0.001231	0.00822	224.6	388.7	0.6555	1.0665
401.0	3494.9	0.001245	0.00787	228.2	387.5	0.6643	1.0616
402.7	3597.6	0.001260	0.00751	231.9	386.1	0.6732	1.0562
404.4	3702.3	0.001275	0.00717	235.6	384.6	0.6821	1.0505
406.1	3809.1	0.001292	0.006835	239.4	382.8	0.6910	1.0443
407.8	3918.1	0.001310	0.006503	243.2	380.9	0.7001	1.0377
409.5	4029.3	0.001330	0.006175	247.1	378.6	0.7093	1.0305
411.2	4142.7	0.001351	0.005850	251.1	376.2	0.7188	1.0228
412.9	4258.3	0.001375	0.005525	255.3	373.4	0.7286	1.0145
414.6	4376.1	0.001402	0.005199	259.7	370.3	0.7388	1.0055
416.4	4496.2	0.001432	0.004868	264.4	366.8	0.7498	0.9956
418.1	4618.5	0.001468	0.004527	269.5	362.7	0.7616	0.9845
419.8	4743.1	0.001512	0.004168	275.3	358.0	0.7749	0.9720
421.5	4870.1	0.001571	0.003775	282.0	352.2	0.7903	0.9569
423.2	4999.4	0.001661	0.003307	290.5	344.1	0.8099	0.9365
424.9	5131.0	0.002315	0.002315	323.0	323.0	0.8859	0.8859

Appendix C. Redlich-Kwong-Soave EoS

Redlich and Kwong (RK) published one of the first successful equations of state and also a methodology for handling mixtures.[34] Soave proposed a modification (RKS) to the original equation, which significantly improved its accuracy.[35] Soave's modification incorporates the Pitzer acentric factor.[36] The original RK equation of state (EoS) is:

$$P = \frac{RT}{(V-b)} - \frac{a}{V(V+b)\sqrt{T}} \tag{C.1}$$

where R is the ideal gas constant. The constants a and b are somewhat empirical and selected to represent the fluid:

$$a = 0.42748\frac{R^2 T_C^{2.5}}{P_C} \tag{C.2}$$

$$b = 0.08664\frac{RT_C}{P_C} \tag{C.3}$$

Equation C.1 can be nondimensionalized by substituting:

$$A = \frac{aZ_C}{P_C V_C^2 \sqrt{T_C}} \tag{C.4}$$

$$B = \frac{b}{V_C} \tag{C.5}$$

to become:

$$Z_C P_R = \frac{T_R}{(V_R - B)} - \frac{A}{V_R(V_R + B)\sqrt{T_R}} \tag{C.6}$$

where $P_R = P/P_C$ is the reduced pressure, $T_R = T/T_C$ is the reduced temperature, and $V_R = V/V_R$ is the reduced specific volume. The critical pressure is P_C, temperature T_C, specific volume V_C, and compressibility $Z = PV/RT$. Compressibility at the critical point is $Z_C = P_C V_C/RT_C$. The critical compressibility is fixed at one-third, $Z_C = 1/3$.

---

[34] Redlich, O. and Kwong, J. N. S., "On The Thermodynamics of Solutions," *Chemical Review*, Vol. 44, No. 1, pp. 233–244, 1949.

[35] Soave, G., "Equilibrium Constants from a Modified Redlich-Kwong Equation of State," *Chemical Engineering Science*, Vol. 27, No. 6, pp. 1197–1203, 1972.

[36] Pitzer, K. S., et al., "Volumetric and Thermodynamic Properties of Fluids II: Compressibility Factor, Vapor Pressure, and Entropy of Vaporization," *Journal of the American Chemical Society*, Vol. 77, pp. 3433, 1955.

Soave's modification replaces the constant A with:

$$A = 0.42748 \left[1 + \left(0.480 + 1.574w - 0.176w^2 \right)\left(1 - \sqrt{T_R} \right) \right]^2 \qquad (C.7)$$

where w is the Pitzer acentric factor:

$$w = -\log_{10}\left(P_R^{sat} \right) - 1$$
$$at\, T_R = 0.7 \qquad (C.8)$$

The RKS EoS works fairly well for gases, but liquid volumes are too large and the subcooled (below saturated liquid) isotherms are too steep (green curves on the left side of the figure below). These same deficiencies are true of all equations of state containing the term $1/(V_R\text{-}B)$.

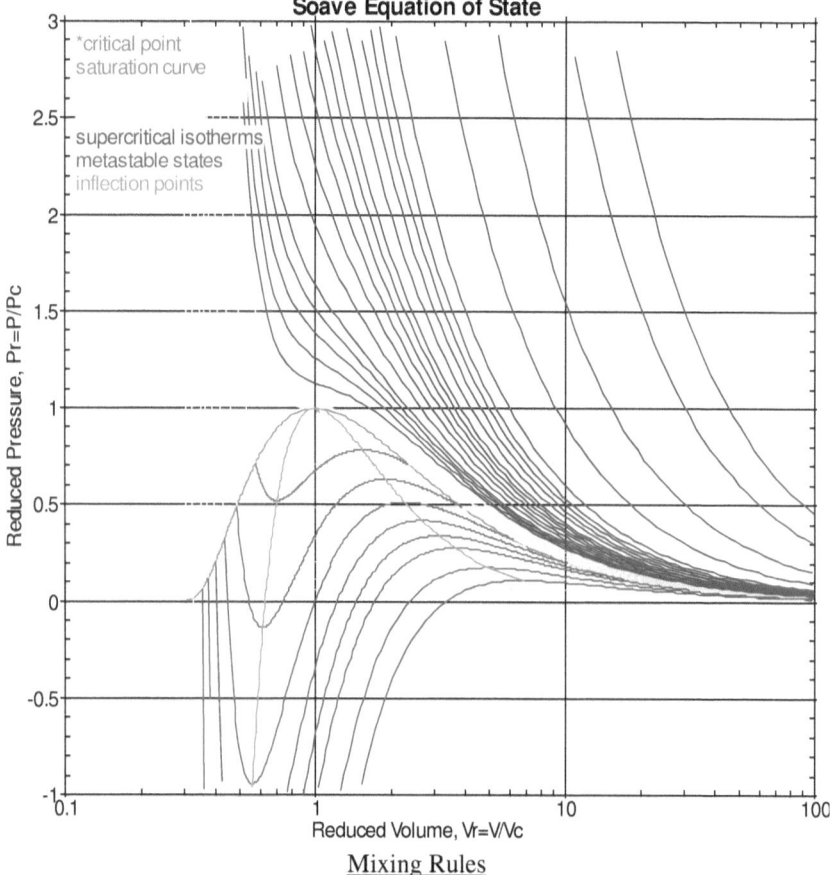

Mixing Rules

Aside from improving upon the van der Waals EoS, the greatest contribution of Redlich and Kwong is in the area of mixing rules and their comparison to measured data, hence the title of the paper. The modified parameters for a mixture are:

$$a_{mix} = \sum_i \sum_j y_i y_j \sqrt{a_i a_j} \qquad (C.9)$$

$$b_{mix} = \sum_i y_i b_i \qquad (C.10)$$

While these rules work fairly well for some mixtures, they do not predict azeotropic behavior because they don't contain enough information about the molecular interactions.

Residual Enthalpy and Entropy

We are interested in properties beyond pressure, temperature, and specific volume or density. We also need enthalpy and entropy. We derive these from Maxwell's relationships. First, the dependence of entropy on specific volume:

$$\left(\frac{\partial S}{\partial V} \right)_T = \left(\frac{\partial P}{\partial T} \right)_V \qquad (C.11)$$

We add two auxiliary relationships, which relate enthalpy to pressure, temperature, and specific volume:

$$\left(\frac{\partial H}{\partial S} \right)_P = T \qquad (C.12)$$

$$\left(\frac{\partial H}{\partial P} \right)_S = V \qquad (C.13)$$

In the ideal gas state (unity compressibility, $Z=1$), there is no dependence on pressure for either entropy or enthalpy; therefore, we can calculate the ideal gas properties at any temperature from specific heat of the dilute (rarefied, vanishing density, infinitesimal pressure) gas. The premier reference for these ideal gas specific heats is the NASA Glenn Report.[37]

$$H_0 = \int_{T_0}^T C_P dT \qquad (C.14)$$

$$S_0 = \int_{T_0}^T \frac{C_P}{T} dT \qquad (C.15)$$

Enthalpy and entropy are always relative to some state. The temperature of the reference state in Equations C.14 and C.15 is T_0. To find the enthalpy or entropy at some finite pressure, we simply integrate Equations C.11 through

---

[37] McBride, B. J., Zehe, M. J., Gordon, S., "NASA Glenn Coefficients for Calculating Thermodynamic Properties of Individual Species," NASA Report No. 211556, 2002.

C.13 from zero to the desired pressure or from the desired specific volume to infinity (zero pressure). The differences between the actual and ideal gas enthalpy and entropy are called *residual* enthalpy and *residual* entropy, respectively. It is convenient to normalize these, dividing residual entropy by the ideal gas constant and residual enthalpy by the ideal gas constant and the critical temperature, as these transform variables yield interesting plots.

$$\frac{S_0 - S}{R} = -\ln Z + \int_\infty^{V_R} \left(\frac{1}{V_R} - \frac{\partial P_R}{\partial V_R} \right) dV_R \qquad (C.16)$$

$$\frac{H_0 - H}{RT_C} = T_R(1 - Z) - Z_C \int_\infty^{V_R} \left(P_R - T_R \frac{\partial P_R}{\partial T_R} \right) dV_R \qquad (C.17)$$

Equations C.16 and C.17 are most often presented in textbooks as integrals with respect to pressure, rather than specific volume. While these are interesting from a historical point of view, they are useless for computation, as there are no equations of state for V in terms of P. Equations of state (such as C.1) are for P in terms of V; thus, the integrals must be over V and not P. It is also more convenient computationally to introduce another equation:

$$\frac{S_0 - S}{R} = \frac{H_0 - H}{RT_C T} + \ln F \qquad (C.18)$$

In the preceding equation, F, is the fugacity coefficient, or the ratio of the ideal to actual pressure. In this case, the ideal pressure is that which would make the ideal gas entropy equal to the actual entropy. This is also defined by an integral:

$$\ln F = Z - 1 - \ln Z - \int_\infty^V \left(\frac{P}{RT} - \frac{1}{V} \right) dV \qquad (C.19)$$

We can integrate Equation C.1 through these steps to arrive at the following relationships, which you will find in the form of code (RKS.c) in the online archive in folder examples.

$$\ln F = Z - 1 - \ln Z + \ln\left(\frac{V_R}{V_R - B} \right) - \frac{A}{BT_R} \ln\left(\frac{V_R + B}{V_R} \right) \qquad (C.20)$$

$$\frac{H_0 - H}{RT_C} = T_R(1 - Z) + \left(\frac{A}{B} - \frac{T_R}{B} \frac{dA}{dT_R} \right) \ln\left(\frac{V_R + B}{V_R} \right) \qquad (C.21)$$

The RKS compressibility is shown in this next figure. Colors are the same as the previous P-V chart. The vapor dome is the red curve with the critical point indicated by a red asterisk.

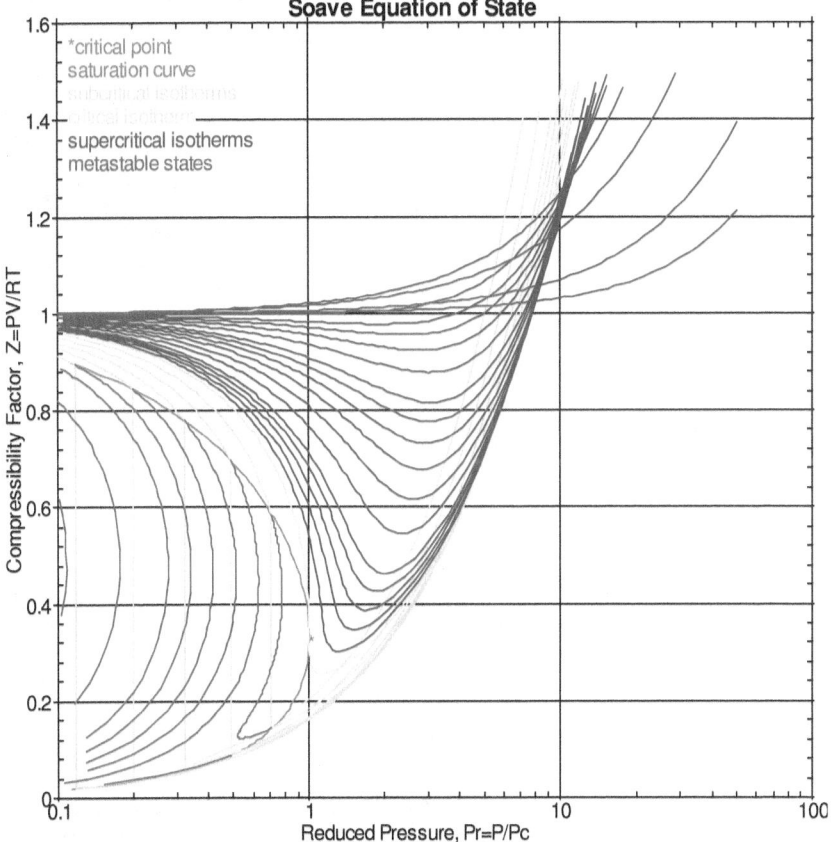

The RKS fugacity coefficient is shown in the figure below:

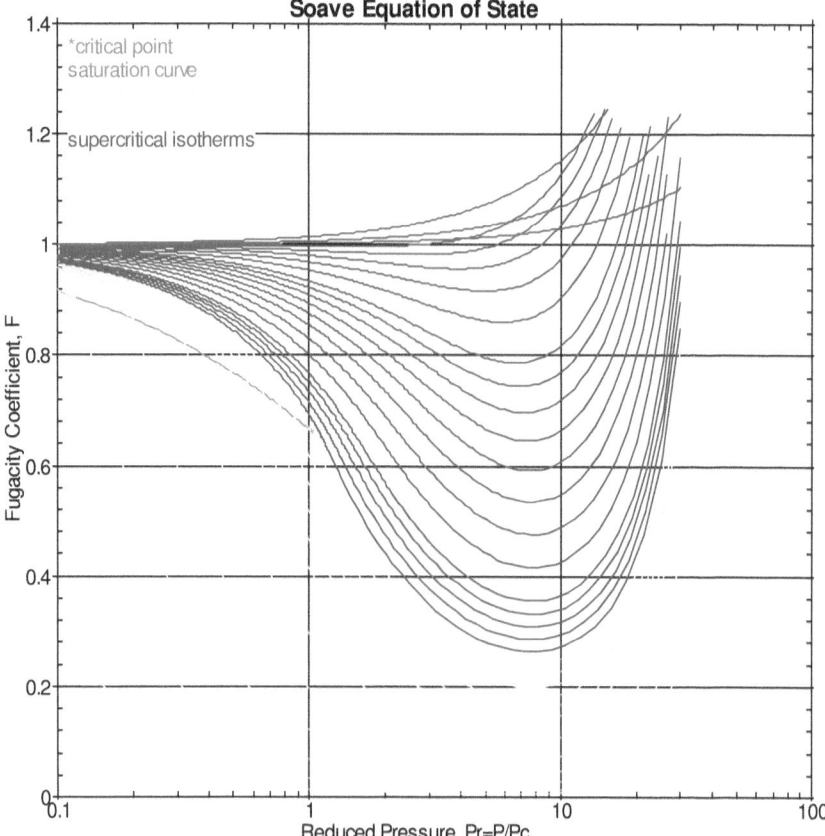

Note that the vapor dome reduces to a single curve on this figure, as the free energy of the saturated liquid and vapor are the same; thus, the fugacity (and fugacity coefficient) must also be equal.

The RKS residual enthalpy is shown in this next figure:

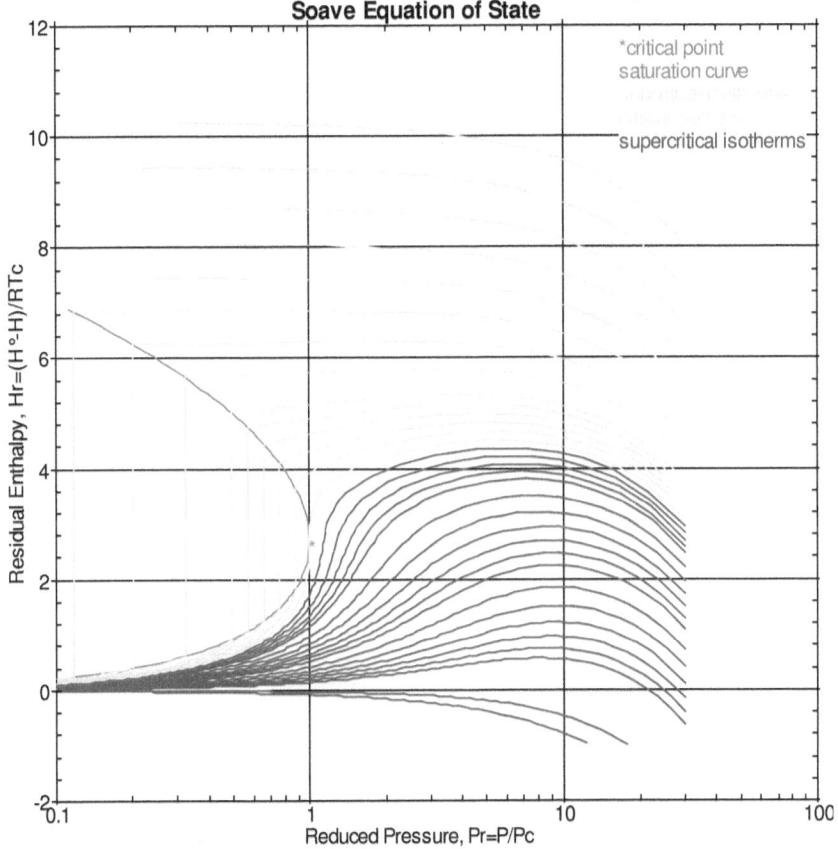

The residual entropy is shown in the figure below:

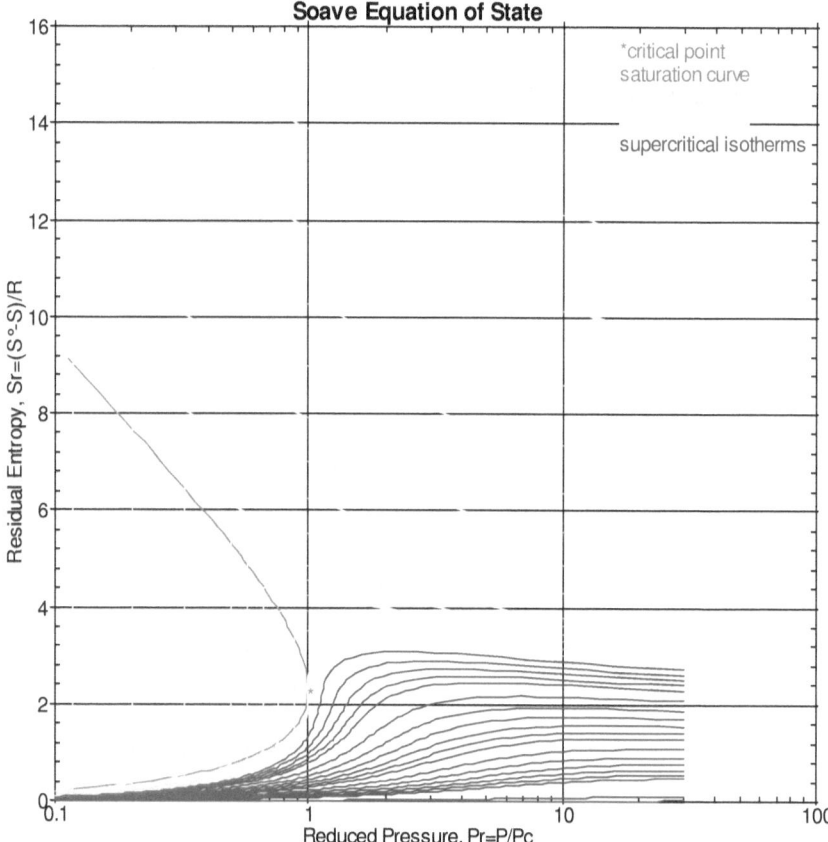

Maxwell's Criterion follows directly from the fact that the Gibbs free energy (g=h-Ts) of the saturated liquid and vapor must be equal for molecules to pass back and forth across the interface. It can be expressed as follows:

$$\int_{V_F}^{V_G} P dV = P_{SAT}\left(V_G - V_F\right) \qquad \text{(C.22)}$$

Applying this integral to Equation C.6 we get:

$$Z_C P_{RSAT}\left(V_{RG} - V_{RF}\right) = T_R \ln\left(\frac{V_{RG} - B}{V_{RF} - B}\right)$$
$$- \frac{A}{B}\left[\ln\left(\frac{V_{RG}}{V_{RF}}\right) + \ln\left(\frac{V_{RF} + B}{V_{RG} + B}\right)\right] \qquad \text{(C.23)}$$

Appendix D. Benedict-Webb-Rubin EoS

The original Benedict-Webb-Rubin[38] (BWR) equation of state was introduced in 1940 and had eight adjustable parameters:

$$Z_C P_R = \frac{T_R}{V_R} + \frac{aT_R}{V_R^2} + \frac{cT_R}{V_R^3} - \frac{d}{V_R^2} - \frac{e}{V_R^3} + \frac{f}{V_R^6} - \frac{h}{T_R^2 V_R^2}$$

$$+ \frac{h\left(1 + \dfrac{g}{V_R^2}\right)}{T_R^2 V_R^3 e^{\left(\frac{g}{V_R^2}\right)}} \tag{D.1}$$

This form has limited accuracy. For the Excel AddIn, we use the 32-parameter extended version presented by Starling[39] in 1973.

$$P = \sum_{i=1}^{9} a_i(T)\rho^i + \exp(-\delta^2) \sum_{i=10}^{15} a_i(T)\rho^{2i-17}$$

$$a_1 = RT, \qquad \delta = \rho/\rho_c$$
$$a_2 = b_1 T + b_2 T^{0.5} + b_3 + b_4/T + b_5/T^2,$$
$$a_3 = b_6 T + b_7 + b_8/T + b_9/T^2,$$
$$a_4 = b_{10}T + b_{11} + b_{12}/T, \quad a_{10} = b_{20}/T^2 + b_{21}/T^3,$$
$$a_5 = b_{13}, \qquad\qquad a_{11} = b_{22}/T^2 + b_{23}/T^4,$$
$$a_6 = b_{14}/T + b_{15}/T^2, \qquad a_{12} = b_{24}/T^2 + b_{25}/T^3,$$
$$a_7 = b_{16}/T, \qquad\qquad a_{13} = b_{26}/T^2 + b_{27}/T^4,$$
$$a_8 = b_{17}/T + b_{18}/T^2, \qquad a_{14} = b_{28}/T^2 + b_{29}/T^3,$$
$$a_9 = b_{19}/T^2, \qquad\qquad a_{15} = b_{30}/T^2 + b_{31}/T^3 + b_{32}/T^4 \tag{D.2}$$

The PVT behavior of the 39 refrigerants are taken from the saturation tables (T_{SAT}, P_{SAT}, V_F, V_G) gathered from various sources. These values are augmented with compressed liquid values based on the isothermal compressibility and the saturated liquid densities after McGowan.[40] Further values are added along the critical isotherm to force the correct density dependence based on the work of

[38] Benedict, M., Webb, G. B. and Rubin, L. C., "An Empirical Equation for Thermodynamic Properties of Light Hydrocarbons and Their Mixtures," *Journal of Chemical Physics*, Vol. 8, pp. 334-345, 1940.

[39] Starling, K. E., *Fluid Properties for Light Petroleum Systems*, Gulf Publishing Company, 1973.

[40] McGowan, J. C., "Variation of the Isothermal Compressibilities of Liquids with Temperature," *Nature*, Vol. 210, No. 5042, pp. 1255-1256, 1966.

Larsen and Sengers.[41] The zero-density (vanishing pressure) specific heats were supplied by the NASA Glenn Report. These parameters were collected for all 39 fluids and subjected to linear constrained minimization to arrive at the best coefficients. The exponential parameter, g, was manually adjusted for each fluid so as to have the desired subcritical isotherms and appropriate number of saturation line crossings (three) in order to assure Maxwell's Criterion was met. For example, the following PV isotherms are ideal:

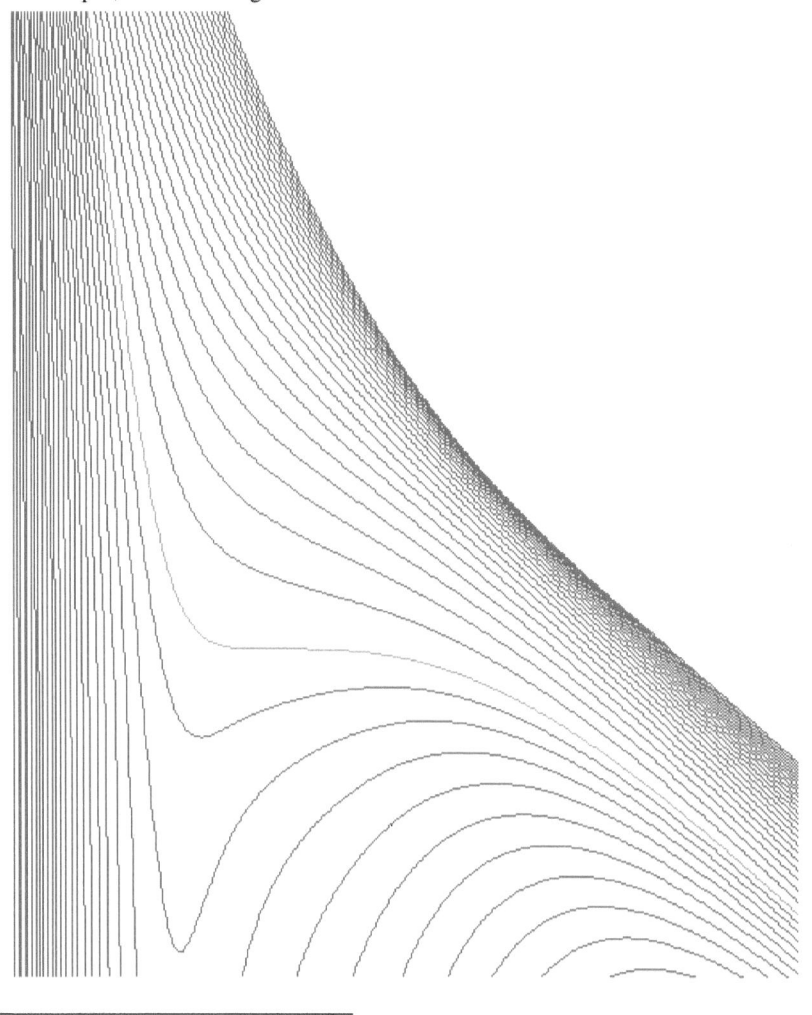

[41] Larsen, S. Y. and Sengers, J. M. H., "On the Behavior of the Compressibility along the Critical Isotherm," *Advances in Thermophysical Properties at Extreme Temperatures and Pressures, Third Symposium on Thermophysical Properties*, Heat Transfer Division of the American Society of Mechanical Engineers, pp. 74-75, 1965.

The following PV isotherms are unacceptable:

Enthalpy and entropy are derived using Equations C.14 through C.19, using Equation D.2 for pressure. The AddIn (Refrig32.xll or Refrig64.xll) will compute a variety of properties, as listed at the beginning of Chapter 5, including saturation properties illustrated in the following figure:

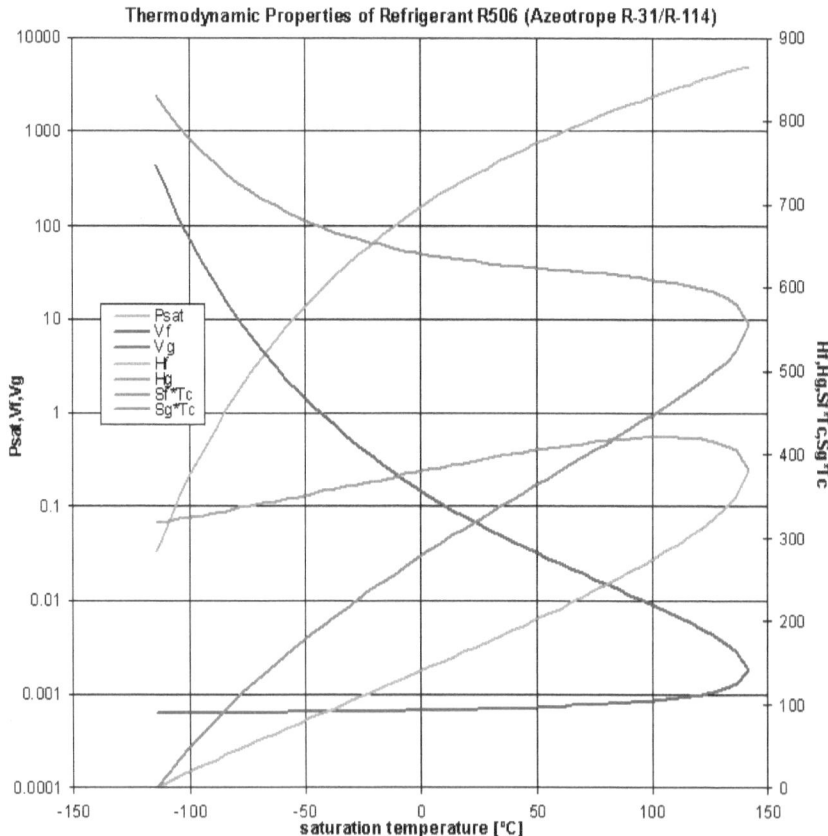

Thermodynamic Properties of Refrigerant R506 (Azeotrope R-31/R-114)

Compressibility is shown in this next figure:

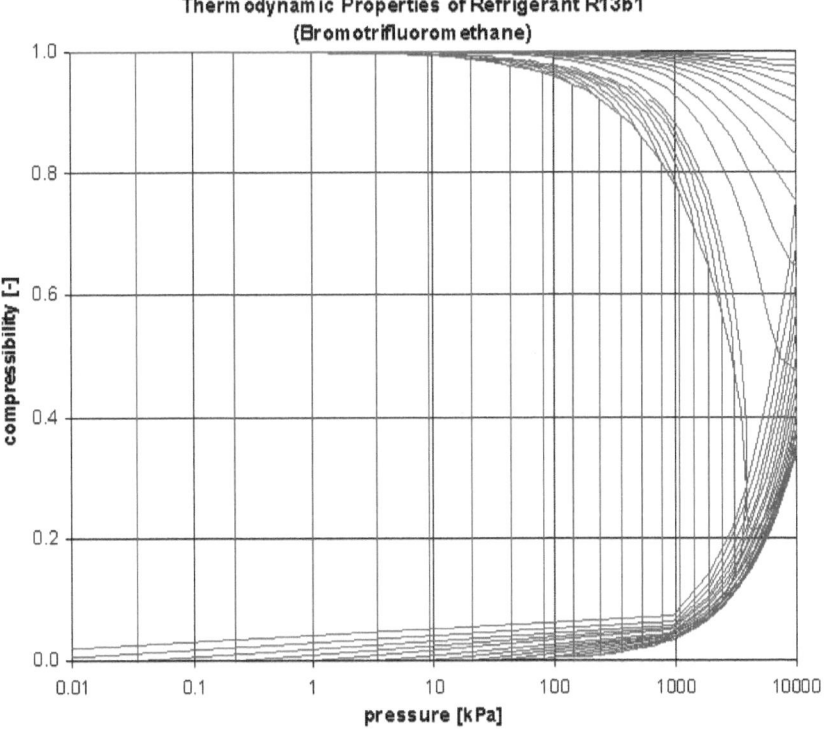

Thermodynamic Properties of Refrigerant R13b1
(Bromotrifluoromethane)

Pressure vs. enthalpy is shown in the figure below:

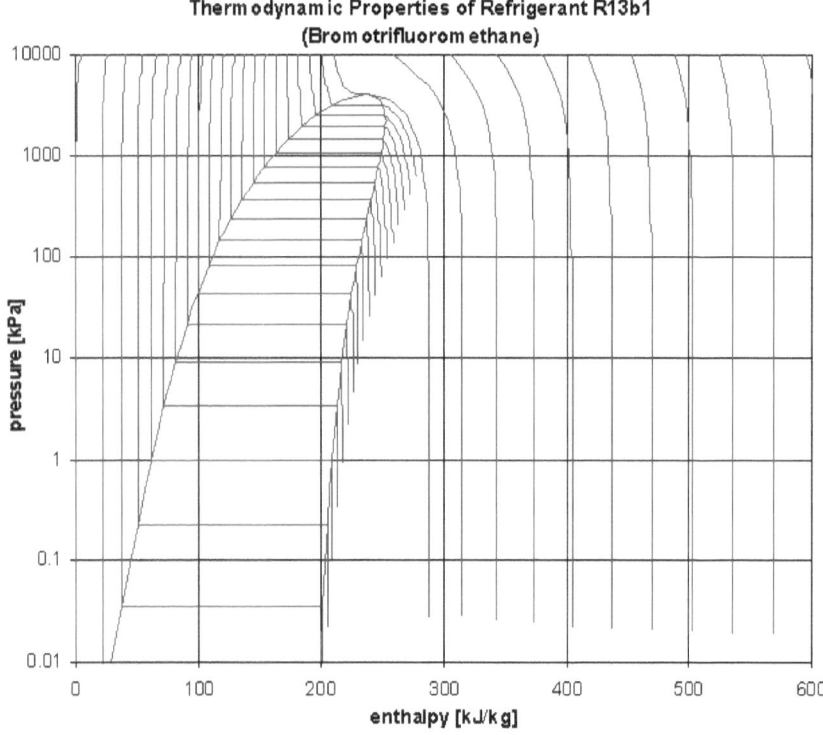

Temperature vs. entropy is shown in the figure below:

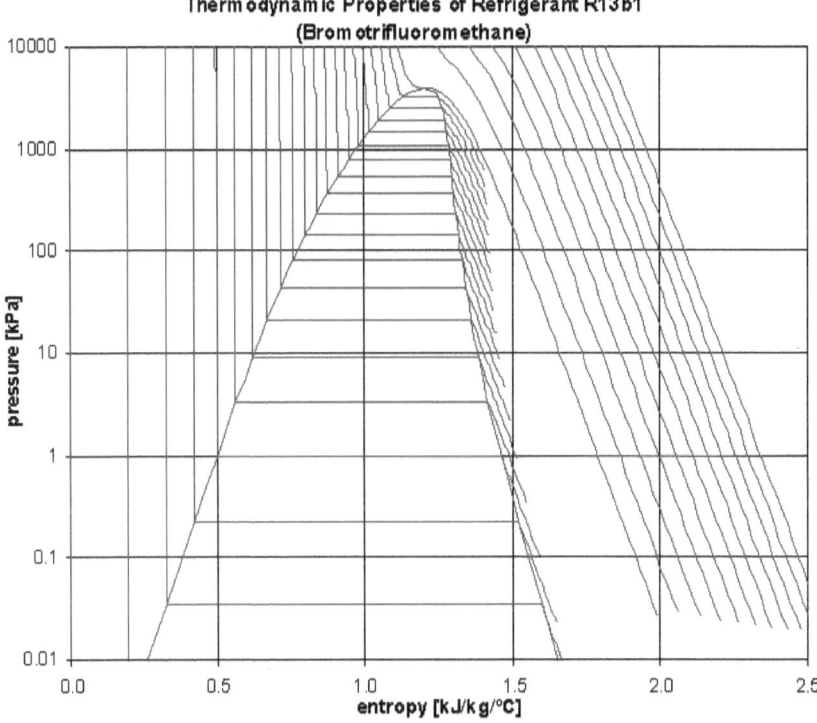

**Thermodynamic Properties of Refrigerant R13b1
(Bromotrifluoromethane)**

An enthalpy vs. entropy (Mollier) diagram, which you will find in the spreadsheet, along with the necessary functions, is shown below:

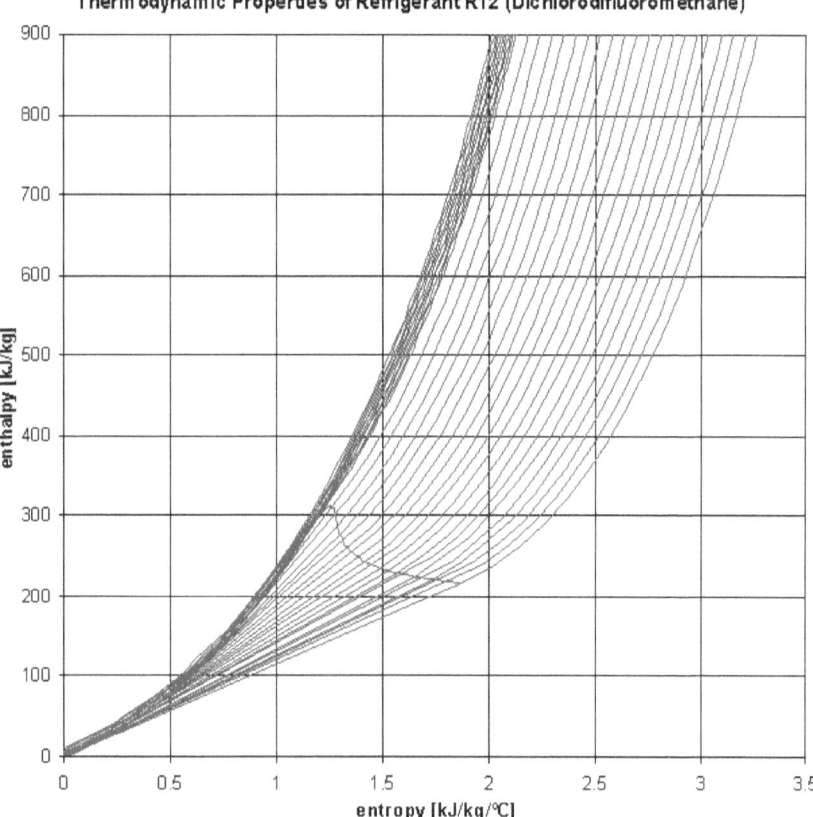

Thermodynamic Properties of Refrigerant R12 (Dichlorodifluoromethane)

Implicit functions are provided so that you can also plot isobars as well as isotherms, as illustrated in this next figure:

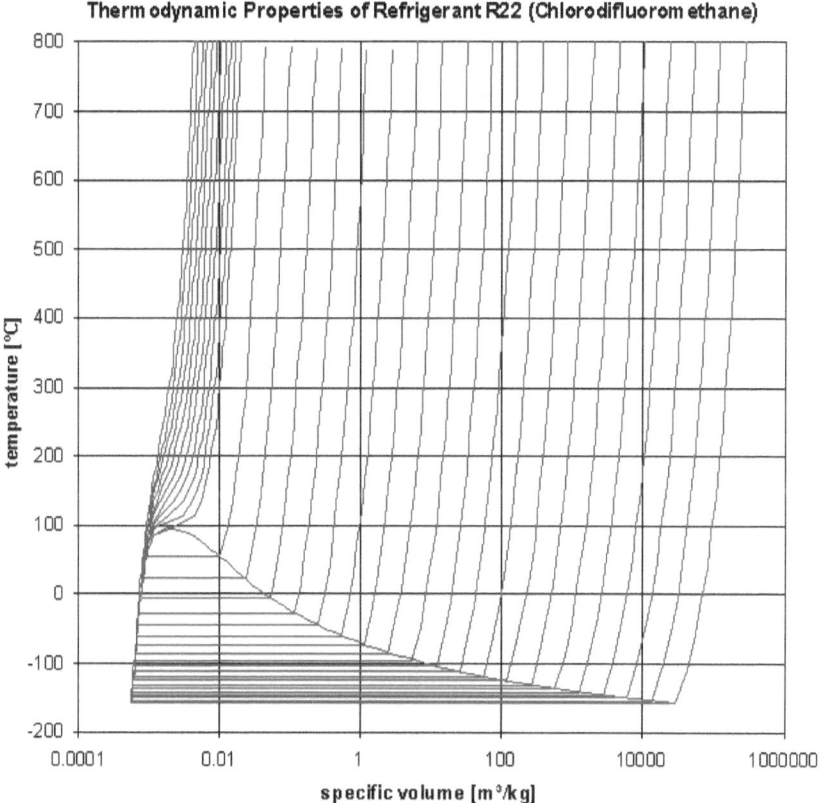

Thermodynamic Properties of Refrigerant R22 (Chlorodifluoromethane)

also by D. James Benton

3D Articulation: Using OpenGL, ISBN-9798596362480, Amazon, 2021 (book 3 in the 3D series).

3D Models in Motion Using OpenGL, ISBN-9798652987701, Amazon, 2020 (book 2 in the 3D series.

3D Rendering in Windows: How to display three-dimensional objects in Windows with and without OpenGL, ISBN-9781520339610, Amazon, 2016 (book 1 in the 3D series).

A Synergy of Short Stories: The whole may be greater than the sum of the parts, ISBN-9781520340319, Amazon, 2016.

bat-Elohim: Book 3 in the Little Star Trilogy, ISBN-9781686148682, Amazon, 2019.

Boilers: Performance and Testing, ISBN: 9798789062517, Amazon 2021.

Combined 3D Rendering Series: 3D Rendering in Windows®, 3D Models in Motion, and 3D Articulation, ISBN-9798484417032, Amazon, 2021.

Complex Variables: Practical Applications, ISBN-9781794250437, Amazon, 2019.

Compression & Encryption: Algorithms & Software, ISBN-9781081008826, Amazon, 2019.

Computational Fluid Dynamics: an Overview of Methods, ISBN-9781672393775, Amazon, 2019.

Computer Simulation of Power Systems: Programming Strategies and Practical Examples, ISBN-9781696218184, Amazon, 2019.

Contaminant Transport: A Numerical Approach, ISBN-9798461733216, Amazon, 2021.

CPUnleashed! Tapping Processor Speed, ISBN-9798421420361, Amazon, 2022.

Curve-Fitting: The Science and Art of Approximation, ISBN-9781520339542, Amazon, 2016.

Death by Tie: It was the best of ties. It was the worst of ties. It's what got him killed., ISBN-9798398745931, Amazon, 2023.

Differential Equations: Numerical Methods for Solving, ISBN-9781983004162, Amazon, 2018.

Equations of State: A Graphical Comparison, ISBN-9798843139520, Amazon, 2022.

Evaporative Cooling: The Science of Beating the Heat, ISBN-9781520913346, Amazon, 2017.

Forecasting: Extrapolation and Projection, ISBN-9798394019494, Amazon 2023.

Heat Engines: Thermodynamics, Cycles, & Performance Curves, ISBN-9798486886836, Amazon, 2021.

Heat Exchangers: Performance Prediction & Evaluation, ISBN-9781973589327, Amazon, 2017.

Heat Recovery Steam Generators: Thermal Design and Testing, ISBN-9781691029365, Amazon, 2019.

Heat Transfer: Heat Exchangers, Heat Recovery Steam Generators, & Cooling Towers, ISBN-9798487417831, Amazon, 2021.

Heat Transfer Examples: Practical Problems Solved, ISBN-9798390610763, Amazon, 2023.

The Kick-Start Murders: Visualize revenge, ISBN-9798759083375, Amazon, 2021.

Jamie2: Innocence is easily lost and cannot be restored, ISBN-9781520339375, Amazon, 2016-18.

Kyle Cooper Mysteries: Kick Start, Monte Carlo, and Waterfront Murders, ISBN-9798829365943, Amazon, 2022.

The Last Seraph: Sequel to Little Star, ISBN-9781726802253, Amazon, 2018.

Little Star: God doesn't do things the way we expect Him to. He's better than that! ISBN-9781520338903, Amazon, 2015-17.

Living Math: Seeing mathematics in every day life (and appreciating it more too), ISBN-9781520336992, Amazon, 2016.

Lost Cause: If only history could be changed..., ISBN-9781521173770, Amazon, 2017.

Mass Transfer: Diffusion & Convection, ISBN-9798702403106, Amazon, 2021.

Mill Town Destiny: The Hand of Providence brought them together to rescue the mill, the town, and each other, ISBN-9781520864679, Amazon, 2017.

Monte Carlo Murders: Who Killed Who and Why, ISBN-9798829341848, Amazon, 2022.

Monte Carlo Simulation: The Art of Random Process Characterization, ISBN-9781980577874, Amazon, 2018.

Nonlinear Equations: Numerical Methods for Solving, ISBN-9781717767318, Amazon, 2018.

Numerical Calculus: Differentiation and Integration, ISBN-9781980680901, Amazon, 2018.

Numerical Methods: Nonlinear Equations, Numerical Calculus, & Differential Equations, ISBN-9798486246845, Amazon, 2021.

Orthogonal Functions: The Many Uses of, ISBN-9781719876162, Amazon, 2018.

Overwhelming Evidence: A Pilgrimage, ISBN-9798515642211, Amazon, 2021.

Particle Tracking: Computational Strategies and Diverse Examples, ISBN-9781692512651, Amazon, 2019.

Plumes: Delineation & Transport, ISBN-9781702292771, Amazon, 2019.

Power Plant Performance Curves: for Testing and Dispatch, ISBN-9798640192698, Amazon, 2020.

Practical Linear Algebra: Principles & Software, ISBN-9798860910584, Amazon, 2023.

Props, Fans, & Pumps: Design & Performance, ISBN-9798645391195, Amazon, 2020.

Remediation: Contaminant Transport, Particle Tracking, & Plumes, ISBN-9798485651190, Amazon, 2021.

ROFL: Rolling on the Floor Laughing, ISBN-9781973300007, Amazon, 2017.

Seminole Rain: You don't choose destiny. It chooses you, ISBN-9798668502196, Amazon, 2020.

Septillionth: 1 in 10^{24}, ISBN-9798410762472, Amazon, 2022.

Software Development: Targeted Applications, ISBN-9798850653989, Amazon, 2023.

Software Recipes: Proven Tools, ISBN-9798815229556, Amazon, 2022.

Steam 2020: to 150 GPa and 6000 K, ISBN-9798634643830, Amazon, 2020.

Thermochemical Reactions: Numerical Solutions, ISBN-9781073417872, Amazon, 2019.

Thermodynamic and Transport Properties of Fluids, ISBN-9781092120845, Amazon, 2019.

Thermodynamic Cycles: Effective Modeling Strategies for Software Development, ISBN-9781070934372, Amazon, 2019.

Thermodynamics - Theory & Practice: The science of energy and power, ISBN-9781520339795, Amazon, 2016.

Version-Independent Programming: Code Development Guidelines for the Windows® Operating System, ISBN-9781520339146, Amazon, 2016.

The Waterfront Murders: As you sow, so shall you reap, ISBN-9798611314500, Amazon, 2020.

Weather Data: Where To Get It and How To Process It, ISBN-9798868037894, Amazon, 2023.